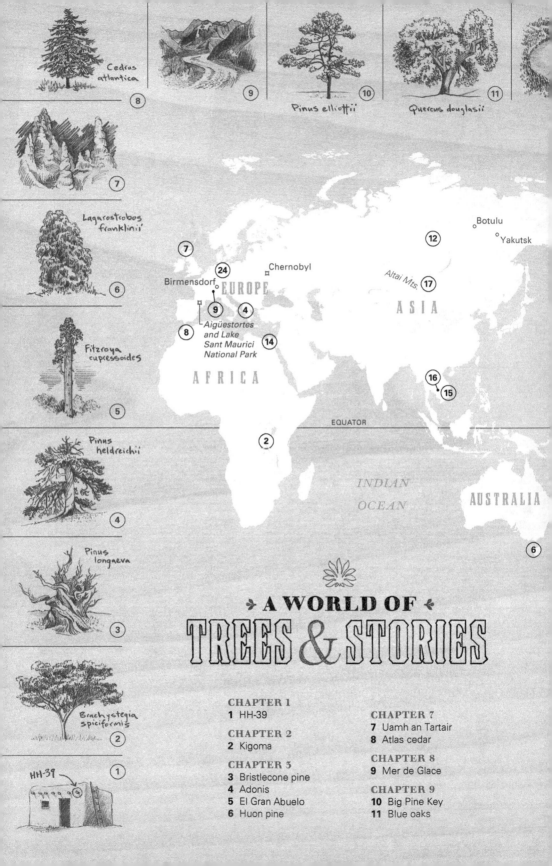

Cedrus atlantica **8**

Pinus elliottii **10**

Quercus douglasii **11**

9

7

Lagarostrobos franklinii' **6**

Fitzroya cupressoides **5**

Pinus heldreichii **4**

Pinus longaeva **3**

Brachystegia spiciformis **2**

HH-39 **1**

Botulu ○
Yakutsk ○

12

Chernobyl □

Altai Mts.

17

24

Birmensdorf ○

EUROPE

9

4

8

Aigüestortes and Lake Sant Maurici National Park

14

A S I A

16
15

AFRICA

EQUATOR

2

INDIAN

OCEAN

AUSTRALIA

6

⇒ A WORLD OF ⇐
TREES & STORIES

Picea sitchensis

Nilometer

Fokienia

Pinus sibirica

ARCTIC OCEAN

Taxodium mucronatum

NORTH AMERICA

Lees Ferry

Tucson

PACIFIC OCEAN

Pueblo Bonito

ATLANTIC OCEAN

SOUTH AMERICA

Taxodium distichum

fire scar

Schöningen spear, oldest wooden artifact > 300,000 years ago

Picea sitchensis

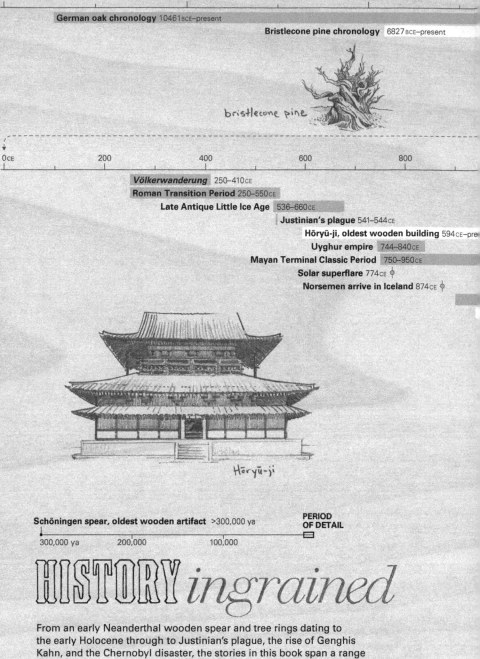

11000 BCE   9000   7000

German oak chronology 10461 BCE–present

Bristlecone pine chronology 6827 BCE–present

bristlecone pine

0 CE   200   400   600   800

Völkerwanderung 250–410 CE

Roman Transition Period 250–550 CE

Late Antique Little Ice Age 536–660 CE

Justinian's plague 541–544 CE

Hōryū-ji, oldest wooden building 594 CE–pre

Uyghur empire 744–840 CE

Mayan Terminal Classic Period 750–950 CE

Solar superflare 774 CE

Norsemen arrive in Iceland 874 CE

Hōryū-ji

Schöningen spear, oldest wooden artifact >300,000 ya

PERIOD OF DETAIL

300,000 ya   200,000   100,000

HISTORY *ingrained*

From an early Neanderthal wooden spear and tree rings dating to
the early Holocene through to Justinian's plague, the rise of Genghis
Kahn, and the Chernobyl disaster, the stories in this book span a range
of global human, climate, and dendrochronology events.

0 CE   200   400   600   800

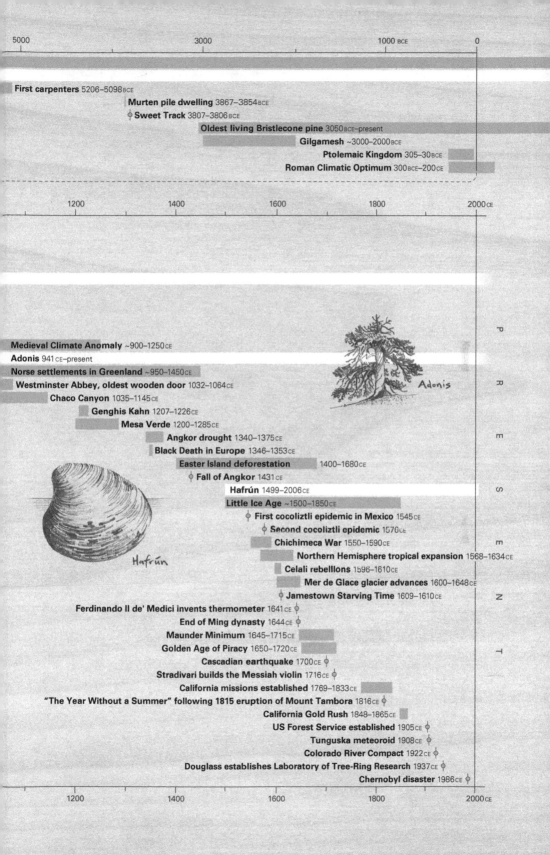

5000   3000   1000 BCE   0

First carpenters 5206–5098BCE

Murten pile dwelling 3867–3854BCE

Sweet Track 3807–3806BCE

Oldest living Bristlecone pine 3050BCE–present

Gilgamesh ~3000–2000BCE

Ptolemaic Kingdom 305–30BCE

Roman Climatic Optimum 300BCE–200CE

1200   1400   1600   1800   2000CE

Medieval Climate Anomaly ~900–1250CE

Adonis 941CE–present

Norse settlements in Greenland ~950–1450CE

Westminster Abbey, oldest wooden door 1032–1064CE

Chaco Canyon 1035–1145CE

Genghis Kahn 1207–1226CE

Mesa Verde 1200–1285CE

Angkor drought 1340–1375CE

Black Death in Europe 1346–1353CE

Easter Island deforestation 1400–1680CE

Fall of Angkor 1431CE

Hafrún 1499–2006CE

Little Ice Age ~1500–1850CE

First cocoliztli epidemic in Mexico 1545CE

Second cocoliztli epidemic 1570CE

Chichimeca War 1550–1590CE

Northern Hemisphere tropical expansion 1568–1634CE

Celali rebellions 1596–1610CE

Mer de Glace glacier advances 1600–1648CE

Jamestown Starving Time 1609–1610CE

Ferdinando II de' Medici invents thermometer 1641CE

End of Ming dynasty 1644CE

Maunder Minimum 1645–1715CE

Golden Age of Piracy 1650–1720CE

Cascadian earthquake 1700CE

Stradivari builds the Messiah violin 1716CE

California missions established 1769–1833CE

"The Year Without a Summer" following 1815 eruption of Mount Tambora 1816CE

California Gold Rush 1848–1865CE

US Forest Service established 1905CE

Tunguska meteoroid 1908CE

Colorado River Compact 1922CE

Douglass establishes Laboratory of Tree-Ring Research 1937CE

Chernobyl disaster 1986CE

Adonis

Hafrún

P R E S E M E N T

1200   1400   1600   1800   2000CE

TREE
STORY

Valerie Trouet The History of the World Written in Rings

TREE STORY

JOHNS HOPKINS UNIVERSITY PRESS

Baltimore

Johns Hopkins Paperback edition, 2022
9 8 7 6 5 4 3 2 1

Johns Hopkins University Press
2715 North Charles Street
Baltimore, Maryland 21218-4363
www.press.jhu.edu

The Library of Congress has cataloged the hardcover edition of this book as follows:

Names: Trouet, Valerie, author.
Title: Tree story : the history of the world written in rings / Valerie Trouet.
Description: Baltimore : Johns Hopkins University Press, 2020. | Includes bibliographical
    references and index.
Identifiers: LCCN 2019031294 | ISBN 9781421437774 (hardcover) | ISBN 9781421437781
    (ebook)
Subjects: LCSH: Dendrochronology. | Tree-rings.
Classification: LCC QK477.2.A6 T76 2020 | DDC 582.16—dc23

LC record available at https://lccn.loc.gov/2019031294

A catalog record for this book is available from the British Library.

ISBN 9781421443744 (paperback)

*Special discounts are available for bulk purchases of this book. For more information, please
contact Special Sales at specialsales@jh.edu.*

To Ursula, Guillian, and John John

Instructions for living a life

Pay attention.

Be astonished.

Tell about it.

—Mary Oliver (1935–2019)

Contents

Prologue

*In 1939, the Ashmolean Museum in Oxford acquires Antonio
Stradivari's legendary violin, the Messiah. One of the most valuable
musical instruments in existence, its estimated value today tops
$20 million. The Messiah was donated to the museum by London's
famed instrument makers and collectors, W. E. Hill & Sons, who had
previously refused a blank check for the violin from car magnate Henry
Ford. The Hills were determined that instead of being stashed away by
a fabulously wealthy private admirer, the Messiah should be visible for
the public to enjoy and future instrument makers to emulate. However,
60 years later, controversy over this remarkable gift suddenly erupted.
In 1999, the Messiah's authenticity was contested by Stewart Pollens,
the associate conservator of musical instruments at New York's
Metropolitan Museum of Art. To bolster their respective arguments,
both Mr. Pollens and the Hill family commissioned dendrochronologists
to date the Messiah.*

Stradivari built the Messiah, arguably his finest work, in 1716, and the violin
remained in his workshop until his death in 1737. In the 1820s, it was sold to
Luigi Tarisio, an Italian collector and dealer who traveled to Paris regularly.
During his visits, Tarisio would boast about the Stradivari masterpiece to Parisian dealers, but he would never bring it with him to show. Legend has it that
this prompted an eminent contemporary French violinist* to proclaim: "Your
violin is like the Messiah—one always waits and he never appears." After Tarisio's death in 1855, the Messiah was bought by one of these Parisian dealers,

---

*His name was Delphin Alard.

Jean-Baptiste Vuillaume, who was a skilled violin maker himself and well known for copying previously constructed instruments. Vuillaume had the Messiah in his possession for more than 30 years, and this is at the heart of the Messiah controversy: is the Ashmolean Messiah the original Stradivari masterpiece, or is it a masterful nineteenth-century copy crafted by the deft hands of Vuillaume?

To answer this question, *dendrochronology*—from the Greek words for "tree," *dendros*, and "time," *chronos*—comes to the rescue. By measuring the widths of the rings in the wood of the Messiah, the violin can be dated. That is to say, we can determine when the tree from which the Messiah was built grew. The most recent rings in the Messiah's wood reveal the earliest date when the instrument possibly could have been built. If the most recent rings measured on the wood of the Messiah postdated 1737, then the tree from which it was built was still growing after Stradivari's death, and the Messiah could not possibly have been built by him. If, on the other hand, the rings predated 1716, the year that Stradivari supposedly built the Messiah, then this would support the violin's authenticity.

Unfortunately, in this case tree-ring dating only added more fuel to the fire of the controversy. The dendrochronologists hired by Pollens put the last datable Messiah ring in 1738, suggesting that the tree was still growing a year after Stradivari's death. The dendrochronologist hired by the Hill family, on the other hand, dated the most recent ring in the violin back to the 1680s, predating the Messiah's 1716 recorded date of manufacture and supporting its authenticity. Both studies, however, were provisional, because they were conducted based only on photographs of the Messiah, not on the violin itself, and neither was ever published in the peer-reviewed scientific literature.

Since this quarrel unfolded, the dendrochronological dating of musical instruments, and particularly string instruments, has grown in popularity and become much more sophisticated. Advances in tree-ring measurement and image-analysis techniques, for instance, allow scientists to work directly on the actual violins. Some tree-ring laboratories, such as that at the University of Hamburg in Germany, have dated thousands of instruments, which has allowed them to build a large database of *reference tree-ring chronologies* against which to compare the Messiah ring-width measurements. Such reference chronologies from a wide range of tree species and geographical areas help not only to precisely date the wood of musical instruments but also to determine their geographical origin, a technique called *dendroprovenancing*.

In 2016, almost two decades after the Pollens-Hill dispute, British dendrochronologist Peter Ratcliff made use of his extensive reference database for Italian string instruments to put an end to the Messiah controversy. Through his research, Ratcliff found that the tree-ring pattern in the Messiah matched the pattern in another Stradivari violin, the 1724 Ex-Wilhelmj,* so well that the wood from both violins could only have come from the same tree. The origin of the Ex-Wilhelmj in Stradivari's workshop is undisputed, and so Ratcliff's dendrochronological work has confirmed the authenticity of the Ashmolean Messiah (we hope) once and for all.

It is spring 1998 and I am pursuing a master's degree in environmental engineering at Ghent University in Belgium. The time has come to choose a research project for my thesis, but I am behind the curve, having just spent a semester in an exchange program in Germany. My classmates have snapped up the most interesting opportunities, especially those involving travel and research abroad. Anxious to secure a project before the summer break, I approach Professor Hans Beeckman, who teaches vegetation ecology and wood anatomy. He suggests that I consider studying tree rings in Tanzania. This is the first time I have ever heard about dendrochronology, but without much hesitation I say yes.

Until then, I had never considered that the rings in trees contained sufficient information to warrant a scientific discipline. But I had a strong desire to work in the developing world and an interest in climate change, and if dendrochronology was an avenue to combine the two while getting a master's degree, then sure, I would study tree rings. Few scientists in our field dreamt of growing up to be dendrochronologists. Most dendrochronology careers, like mine, start randomly, with serendipitous undergraduate or graduate field- or lab-based opportunities that morph over time into full-fledged careers.

---

*The most recent rings of the Ex-Wilhelmj were dated to 1689 on the treble side and 1701 on the bass side of the violin, both dates prior to the 1716 construction of the Messiah.

The first hurdle I encountered in my dendro-career was to persuade my mom that tree-ring research in Africa would be a valuable capstone project for my engineering degree. "Valerie, you are one year away from getting a degree that will open up a world of exciting and lucrative career opportunities. But tree rings? In Africa? How do you plan to make a career out of that?" In retrospect, my mom was presumably mostly concerned about me traipsing through Africa, inexperienced and unprepared, and her fears were probably valid. But 20 years later, having built a successful international dendrochronology career, I can't help but occasionally remind her of that comment.

It was the lab work for my master's project that got me hooked on tree rings. Looking through a microscope at the wood that I had collected in Tanzania was a game changer. Wood is gorgeous, and finding matching tree-ring patterns is like solving a puzzle—it is addictive. I spent hours in the flow of dendrochronology, fully unaware of time passing. When an opportunity opened up to do four more years of dendrochronological research while pursuing a PhD, it did not take me long to decide. My options, at 25 years old, were to either start a 40-year career as an office-bound government employee or become a scientist, getting paid to travel to Africa and solve more tree-ring puzzles. A no-brainer. If you like it, then you should put a ring on it. Writing a PhD dissertation, however, turned out to be a more lengthy, tedious process than I could possibly have imagined. I spent the last year of my studies in my sixth-floor walk-up flat in downtown Brussels, drinking coffee, smoking cigarettes, and looking out over the skyline while I wrote.

I defended my PhD in December 2004 and moved to the United States immediately after for a postdoctoral position in the Geography Department at Pennsylvania State University in State College. I had been to the US only once before, on a visit to New York City. Little did I know that State College is a small town in the middle of Amish farmland, three hours away from the nearest city. When I landed at the tiny State College airport, with two suitcases and a backpack, my adviser, Alan Taylor, picked me up. I asked him to drive through downtown State College so I could see what it was like, but Alan hesitated and instead drove by the Penn State campus. As it turns out, there is no such thing as "downtown State College": there's the campus and then a few streets with shops and bars, and that's it. Moving there straight from cosmopolitan Brussels was quite the culture shock. But I loved working in

Alan's Vegetation Dynamics lab, linking historical fires in California to past climate. I continued to travel for my job, this time to the Sierra Nevada in California, and to feed my appetite for tree rings in the lab.

It was during my time at Penn State that I met Jan Esper, the head of the dendrosciences group at the Swiss Federal Institute for Forest, Snow, and Landscape Research, or Wald, Schnee und Landschaft (WSL), Europe's most prominent tree-ring lab. When Jan offered me a job, I decided that I had spent enough time in the Pennsylvanian boonies and it was time to return to the Old World, to the city of Zurich. At the WSL, I learned how to use tree rings to reconstruct past climate and how to publish papers in top scientific journals with a broad readership, such as *Nature* and *Science*. But after four years in Switzerland, I crossed the pond once again to take up a professorship at the University of Arizona in the Laboratory of Tree-Ring Research (LTRR), the tree-ring mothership, where dendrochronology originated.

At the LTRR, in charge of my own research group for the first time, I have further explored the boundaries of what tree rings can tell us about the climate of the past. With the input of very talented postdocs and graduate students, we use tree rings to look at past climate extremes, such as droughts in California and hurricanes in the Caribbean, and at climatic movements that occur in the upper echelons of the atmosphere, such as the jet stream.

I am a dendroclimatologist: I use tree rings to study the climate of the past and its impacts on ecosystems and human systems. Over the past 20 years, I've spent most of my days thinking, writing, and speaking about past and future climate change. It can be a daunting task. Year after year, we learn more about the climate and about the havoc that our fossil-fuel burning wreaks upon it. About the consequences of such global, man-made climate change to society—heat waves! hurricanes! snowmageddons!—and to ecosystems—forest fires! polar bears! Yet, year after year too little is done at a governmental level to curb our carbon-dioxide emissions and to mitigate the worst effects of man-made climate change. Even after the 2015 Paris Climate Agreement, in which 196 countries committed to undertaking ambitious efforts to combat climate change, not much has improved. Carbon-dioxide emission levels are at an all-time high, and in the US under President Donald J. Trump, the threat of man-made climate change is not only being widely ignored, it is actually being called a lie, "fake news."

By the beginning of 2017 I was tired and frustrated. Tired of the constant barrage of bad climate news. Frustrated by the constant pressure to defend my expertise, my gender, and even the science that I stand for. I decided not to spend my upcoming sabbatical thinking and writing about the doom and gloom of climate change but instead to write stories of the excitement of scientific discovery, about our long and complex human history, and how it has been intertwined with our natural environment and ingrained in the stories of trees.

Dendrochronology lends itself extraordinarily well to this purpose for two main reasons. First, many people are comfortable with the concept behind dendrochronology, having looked at the top of a tree stump as kids and counted the rings. It is a tangible field of science: you can touch the wood with your own hands, you can see the rings with your naked eye. It does not involve obscure nanoparticles or far-flung galaxies. Second, dendrochronology is in a unique position to reveal the interactions between human history and environmental history, because it sits right at the nexus of ecology, climatology, and human history. And this has been the case since the science first emerged, almost 100 years ago, in the American Southwest.

Over the past century, the evolving scientific field of dendrochronology has produced a network of tree-ring data that continues to expand in space and in time. The global tree-ring network now includes the loneliest tree on earth, on Campbell Island in the Antarctic Ocean, more than 170 miles removed from its nearest neighbor. The longest continuous tree-ring record, the German oak-pine chronology, covers the past 12,650 years without skipping a single year. This growing network of tree-ring data has allowed us to tackle increasingly complex research questions. The global-scale network has empowered us to study past climate not only at the earth's surface, where the trees grow, but also higher up in the atmosphere, where surface climate is orchestrated. It has enabled us to study not only past average climate but also its extremes, its heat waves, hurricanes, and wildfires. The year-by-year, ring-by-ring precision of dendrochronology creates a foothold, a mooring, in the study of the complex interactions between human history and climate history. It has allowed us to move on from simplistic, deterministic characterizations of past climate-society relationships to a more holistic understanding that emphasizes the importance of a society's resilience and adaptive capacity.

In *Tree Story*, I aim to tell how dendrochronology evolved from its humble origins to being one of the primary tools used to study the complex interactions between forests, humans, and climate. The journey is far from linear and full of surprises. Its stories unfold from the field's puzzling origins in the Sonoran Desert, where trees are scarce, to the revelations offered by "counting the rings" in archeological and historical wood, to the epic sagas of the climate of past millennia. I focus on natural and fabricated hazards that have been recorded in tree rings (earthquakes, volcanoes, etc.), and I reveal how past climate changes have impacted societies around the globe, including the Roman Empire in Europe, the Mongol Empire in Asia, and the Ancestral Puebloans in the American Southwest.

I cover a lot of ground in *Tree Story*. I talk about wood cells smaller than the diameter of a strand of human hair and about jet-stream winds that circle the entire Northern Hemisphere at the altitude at which airplanes fly. I link the two through stories that involve pirates, Martians, samurai, and Genghis Khan. I write the tree-ring stories that I find fascinating. And threading through all these stories is the narrative history of wood use and deforestation, which has allowed dendrochronologists to study the past and to contribute to ensuring a livable planet in the future. I think there is a place for such stories of discovery in the current climate of mistrust and disinterest in scientific advances. In the best-case scenario, I hope you will feel a little tingle of excitement when learning something new from this book, the same one that helps us scientists keep on keeping on.

One
Trees in the Desert

By translating the story told by tree rings,
we have pushed back the horizons of history.
—Andrew Ellicott Douglass, 1929

In July 2010, I made the curious decision to move from Zurich, Switzerland, to Tucson, Arizona. Tucson was foundering in the aftermath of the 2008 financial crisis, while Zurich remained the strong-beating heart of one of the world's most stable economies. And although an avid snowboarder, I was trading the Swiss Alps for the Sonoran Desert. Yet, the question I was asked most frequently when sharing my decision with friends and family was unrelated to the economy or snowboarding. Instead, everyone wondered why on earth a tree-ring scientist would move to the desert. "Don't you need trees for your research?"

A fair question. After all, I study the rings in long-lived trees to better understand the climate of the past and how it has influenced human systems as well as ecosystems. Intuitively, Switzerland, with its dense forests, mountain climate, and long documented history, seems like a logical place to be a dendrochronologist. Tucson, located in the Sonoran Desert of southern Arizona, not so much. Why, then, is the Laboratory of Tree-Ring Research (LTRR), the world's first and foremost department dedicated to dendrochronology, situated at the University of Arizona (UA) in Tucson, less than a hundred miles north of the Mexican border?

Upon accepting my position in the LTRR, I did not myself know the full story of why it had originated amid the Saguaro cacti and the Gila monsters. I knew that an astronomer, Andrew Ellicott Douglass (1867–1962), had established the LTRR in the 1930s and that it had since been in operation under the UA football stadium, but not much more. It was not until I started teaching Introduction to Dendrochronology as a freshly minted professor that I learned the details of the historical link between Tucson, astronomy, and tree rings.

In addition to a professorial position in the so-called mecca of tree-ring research, my relocation from Zurich to Tucson offered another benefit: the climate. In Zurich, the sun shines fewer than 4 days out of 10 on average. In Tucson, that number is 9 out of 10, which results in an average rainfall of less than 12 inches per year. Hence the Sonoran Desert landscape. This hot and sunny climate with sparse cloud cover and rainfall cannot support forests as natural vegetation; most trees need much more water to grow their hefty trunks. However, it does support a branch of science that depends on clear, cloudless skies: astronomy. And that is the reason why dendrochronology originates in a place without trees: its founder had moved to Tucson in the early twentieth century in search of lucid and steady skies.

The field of astronomy progressed dramatically throughout the nineteenth century, with improved telescopes and newly invented instruments leading to previously unimaginable detailed examination of stars and nebulae and the discovery of new planets and asteroids. The continued advancement of astronomy depended on precise observations, which required not only modern instrumentation and skilled professionals but also stable atmospheric conditions. To find such conditions, astronomers and observatory builders headed to the American West, where the Lick Observatory in central California was the first to be built, in 1888.

Astronomy swiftly evolved into one of the most captivating sciences of the era, firing the imagination of many bright minds as well as affluent amateurs. One wealthy benefactor of late nineteenth-century astronomy was Percival Lowell, a Harvard-educated businessman who was so fascinated by the planet Mars that he decided to spend all his time and money on the planet's study. In 1892, Lowell committed to financing an observatory in the American Southwest dedicated to the investigation of the red planet, to be built in preparation for the 1894 Mars opposition. Roughly every two years, when Earth passes between the sun and Mars, Mars is in opposition. This is the time when the red planet is at its brightest in the sky and is best studied. Lowell hired Douglass, who was working as an astronomer at the Harvard Observatory at the time, to scout for a location in the Southwest and oversee the observatory's planning. Douglass decided on Flagstaff, in northern Arizona, as the optimal location and supervised the construction of the Lowell Observatory, which was completed by late May 1894, just in time for the Mars opposition. He effectively ran the observatory until 1901, when a long-festering astronomical argument with Lowell ended Douglass's tenure there. The reason for their falling out? Martians.

In his study of Mars, Lowell was inspired by the work of Giovanni Schiaparelli, an Italian astronomer who observed a network of long straight lines on Mars's surface during its 1877 opposition and described them as *canali*. Schiaparelli's description left room for ambiguity, because the Italian word *canale* can mean both "gully" (a natural configuration) and "channel" (an artificial construction). It was translated into English as the latter, and the suggestion of canals on Mars galvanized a slew of hypotheses about intelligent life on other planets. Lowell adhered to a theory that the canals on Mars had been built for irrigation purposes by an intelligent, alien civilization in an arid environment, and he committed a substantial portion of his observations at the Lowell Observatory to supporting this theory. He was also devoted to popularizing the idea of life on Mars among the broader public, and his efforts on this front were reinforced by the publication in 1897 of H. G. Wells's influential novel *The War of the Worlds*, in which Martians abandon a dry, dying planet to invade planet Earth.

Such theories of Martian intelligent life became wildly popular with the wider public, but the professional astronomy community expressed stern skepticism toward the idea. Early observations of the canals had been made using relatively low-resolution telescopes and were based on drawings, not photographs, which left ample room for human error and subjectivity. The theory of canals on Mars was therefore considered controversial even before Lowell's involvement, but as Lowell became more outspoken about an intelligent civilization as the only plausible explanation for the Martian canals, so did his scientific opponents. As the principal observer at the Lowell Observatory, Douglass was unavoidably involved in this scientific controversy.

During the 1894 Mars opposition, the one for which the Lowell Observatory had been constructed, Douglass made numerous observations of the shape, atmosphere, and *canali* of Mars, which bolstered Lowell's Martian assumptions. But growing concerns about the legitimacy of these assumptions encouraged Douglass to investigate potential deficiencies of the observatory and the optical illusions they might lead to. For this purpose, he mounted globes and disks ("artificial planets") at various distances from the observatory and studied them through its telescope. He found many of the telescopically observed surface details on these artificial planets, including long straight lines, to be fabricated optical illusions. Through his experiments, he discovered that the lines observed on Mars, its "canals," were among these optical illusions and that Lowell's assumption of an advanced Martian civilization was ultimately a product of self-delusion.

This realization put a strain on Douglass's relationship with his employer. Douglass grew even more disenchanted with Lowell's aspirations after the "Message from Mars" debacle. During a Mars observation in December 1900, Douglass noticed a particularly bright projection and telegraphed this result to Lowell, who without further study forwarded the news of a bright light coming from Mars to his colleagues at Harvard and in Europe. Within days, the European and the American press picked up the story and zealously interpreted it as a message from Mars's inhabitants. Douglass and his fellow astronomers had to spend the following weeks debunking this story and persuading the public that the observed phenomenon was nothing more than a cloud. After this fiasco, Douglass made little effort to conceal his disdain for his supervisor's approach to doing and disseminating science. In a letter to a colleague in March 1901 he wrote, "It appears to me that Mr. Lowell has a strong literary instinct and no scientific instinct." In another letter he wrote, "I fear it will not be possible to turn [Lowell] into a scientific man." Even though he emphasized the confidentiality of these letters to their recipients, it must not have come as a surprise to Douglass when four months later Lowell dismissed him from his observatory.*

Five years later, in 1906, Douglass found a new job as an assistant professor of physics and geography at the University of Arizona in Tucson, which at the time was an institution with 215 students, 26 faculty members, and no astronomy department. At UA, Douglass promoted the development of astronomy in southern Arizona and succeeded in fundraising, building, and directing the Steward Observatory, which was founded in 1923. In addition to this, Douglass revealed himself as a true Renaissance man: he developed the new scien-

---

* A. E. Douglass to William H. Pickering, 8 March 1901, Box 14, Andrew Ellicott Douglass Papers, Special Collections, University of Arizona Library; Douglass to William L. Putnam, 12 March 1901, Box 16, ibid.

tific field of dendrochronology and made great strides not only in astronomy but also in the fields of paleoclimatology and archeology.

Douglass began the work of dendrochronology in Arizona, where his ambitions in astronomy had taken him. He collected his first 25 tree-ring samples from a log yard in Flagstaff by cutting stem discs from the ends of logs and the tops of stumps. He was motivated to do so by his hypothesis that the rings in trees could be used as tracers of past cycles in the activity of the sun. As an astronomer, Douglass had developed a keen interest in such solar cycles and their influence on Earth's climate and had closely followed the recent developments in this field. These included (1) the identification of an 11-year cycle in the occurrence of sunspots (dark, cool areas on the surface of the sun that are visible through a telescope); (2) the relation of this cycle to a similar cycle in energy coming from the sun; and (3) its potential to influence and create cyclicity in the climate on Earth. For instance, the nineteenth-century English astronomer Norman Lockyer—who also established the scientific journal *Nature* and later married the suffragette Mary Brodhurst—hypothesized a link between the sunspot cycles and Indian monsoon rainfall, a research topic that is still being investigated today, more than a century later. Because of the complexity of both the sun's energy and the earth's climate, long *time series*, sequences of data recorded at successive points in time and listed chronologically, were needed to decipher their relationships. It was Douglass's idea that the annual rings in long-lived trees' trunks could provide such time series.

Douglass reasoned that a tree's growth year after year could be measured by the width of its annual rings. How much the girth of a tree increases each year, and thus the width of its rings, is determined by the tree's food supply. In the American Southwest, as in most semi-arid regions, a tree's food supply depends largely on how much water the tree receives through precipitation generated by snow- and rainfall. Putting one and one together, Douglass hypothesized that a tree's ring width for a certain year is a likely indicator of the amount of rainfall it received in that year. If a connection existed between that amount of rainfall and the amount of energy coming from the sun, as, for instance, in Lockyer's hypothesis, then tree rings could be used as records of past variations not only in rainfall but also, potentially, in solar activity. *Tree-ring series*, time series based on tree-ring data from old trees, could then provide centuries' worth of solar-variation data. Exploring this idea, Douglass collected more than a hundred ponderosa pine (*Pinus ponderosa*) samples in northern Arizona. By 1915 he had developed a *tree-ring chronology** that extended back to 1463 CE,† allowing him to study 450 years' worth of cyclic variations in the growth of the trees.

In search of even older tree-ring material, Douglass traveled further afield to the giant sequoia (*Sequoiadendron giganteum*) groves in the Sierra Nevada in California, whose ancient character had long been recognized. His investigations revealed that the oldest ring in the oldest giant sequoia sample he had collected had been formed in 1305 BCE—he had sampled a tree that was more than 3,200 years old! Because the ponderosa pine trees in northern Arizona never reached such old ages—only two of the trees he sampled there were more than 500 years old—Douglass would need another source of tree-ring material to extend the Arizona record further back in time. It wasn't long before this source presented itself in the form of wood samples extracted from the abundant archeological sites in the Four Corners region of Colorado, New Mexico, Arizona, and Utah.

▨ Southwestern archeology experienced its heyday in sync with Douglass's advancements in dendrochronology. Many of the Ancestral Puebloan ruins and cliff dwellings in the Four Corners region that nowadays are preserved in national monuments and national parks—Chaco Canyon, Mesa Verde, Canyon de Chelly, Casa Grande, and Aztec Ruins—were excavated in the late 1800s and early 1900s. While the discovery of these impressive prehistoric structures captured the public's imagination, it left archeologists with more questions than answers regarding construction and abandonment dates. Early twentieth-century southwestern archeologists used styles of pottery, the most distinctive and abundant artifact across the region, to derive relative dates and to answer questions such as whether Chaco Canyon was built before or after Aztec Ruins. Yet absolute dates remained elusive.

---

* We generally use *tree-ring series* when talking about tree-ring data derived from a single sample and *tree-ring chronology* when talking about crossdated tree-ring data from multiple trees or sites.
† I use CE ("of the Common Era") and BCE ("before the Common Era") throughout this book. These are equivalent to AD (*anno Domini*, "in the year of the Lord") and BC ("before Christ"); for example, "1463 CE" corresponds to "AD 1463," and "1305 BCE" corresponds to "1305 BC."

The American Museum of Natural History in New York was one of the institutes pursuing absolute dates for the Four Corners prehistoric ruins. After reading about Douglass's dendrochronological work, the curator of anthropology at the museum wrote to Douglass: "Your work suggests to me a possible help in the archaeological investigation of the Southwest. . . . We do not know how old these ruins are, but I should be glad to have an opinion from you as to whether it might be possible to connect up with your modern and dated trees specimens [wood specimens] from these ruins by correlating the curves of growth."[*] As a result of this letter, Douglass started collaborating with archeologists as early as 1915, 11 years after collecting his first tree-ring sample in Flagstaff. Douglass's aim was to determine whether he could link the tree-ring patterns in the archeological wood samples from the Four Corners region to the patterns in his 450-year-long living-tree chronology from northern Arizona. If he could find overlap between the two sets of tree-ring patterns, then he could apply the living trees' dates to the archeological wood with annual precision, a big step up from the vaguer date ranges provided by prevailing archeological dating techniques at the time.

Douglass was soon able to date tree-ring series from archeological wooden beams and charcoal[†] from different Four Corners sites relative to each other, but none of the series overlapped in time with his living-tree chronology. As a result, the archeological wood chronology remained *floating*: it could not be linked to or anchored in the present, and precise, or *absolute* dates for the Puebloan ruins were still lacking. But even though it did not yet provide absolute dates, with its potential for relative dating alone, the floating chronology was a uniquely valuable tool. The application of dendrochronology to southwestern archeology led to relative dates for a steadily growing number of sites and to the first accurate chronological ordering of the prehistoric sites in the Four Corners region. Douglass's tree-ring work established, for instance, that all five major ruins at Chaco Canyon had been built within a 20-year timespan and that its Pueblo Bonito complex had been built 40 to 45 years prior to Aztec Ruins.

Douglass's tree-ring contribution to archeology swiftly became invaluable, yet it took him an additional 14 years to grasp the holy grail of absolute dating.

---

[*] Clark Wissler to A. E. Douglass, 22 May 1914.

[†] Charcoal is typically better preserved than regular wood and can show clear tree rings. Large charcoal fragments with a sufficient number of tree rings can be tree-ring dated.

To provide an exact calendar year for the construction of the Ancestral Puebloan buildings, Douglass needed to find the missing link between the floating, archeological tree-ring chronology and the absolutely dated living-tree chronology, which was anchored in his own time (fig. 1). Douglass approached the challenge of bridging the gap between the two time series from two directions: by extending the living-tree chronology as far back in time as possible and by

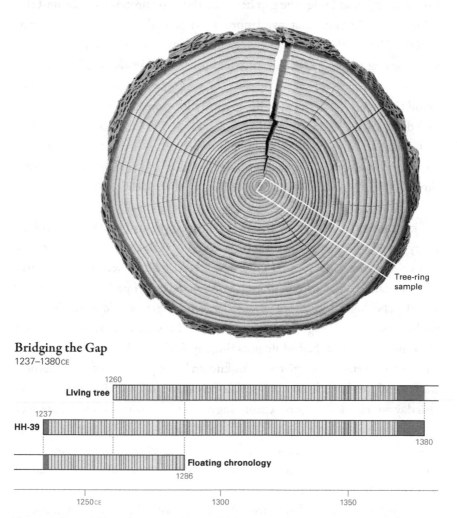

Figure 1 Beam HH-39 showed 143 rings, which overlapped in time with the 120 earliest years of the dated living-tree chronology (1260–1380). Douglass dated its innermost ring to 1237. HH-39 also overlapped with the final 49 years of the floating chronology, creating a firm, continuous record back to 700 CE.

bringing the floating chronology as far forward toward the present as possible. In doing so, he hoped to eventually reach the point when the two chronologies overlapped, when the gap was bridged. By 1929 he had extended the dated living-tree chronology back to 1260 CE. His floating chronology, which included material from 75 Four Corners ruins, spanned 585 years. It took one beam, sampled at a ruin in Show Low, in eastern Arizona, to provide the breakthrough and bridge the gap between the two chronologies. Beam HH-39 showed 143 rings that overlapped in time with the 120 earliest years (1260–1380) of the dated living-tree chronology, and Douglass dated its innermost ring to 1237. He then discovered that HH-39 also overlapped with the most recent 49 years of the floating chronology, whose most recent year he could therefore firmly date as 1286. In one fell swoop, HH-39 unlocked absolute dates for the beams from Show Low (1174–1383), Aztec Ruins (1110–21), Pueblo Bonito (919–1127), and all other ruins that had contributed to the (now no longer) floating chronology. Within a year, Douglass's "Rosetta Stone," HH-39, provided an accurate historical perspective for 75 Ancestral Puebloan ruins.

I am reminded of the magnitude of Douglass's contribution to archeology every time I visit a museum displaying artifacts with age estimates that span centuries or even millennia. As a tree-ring scientist, I have become accustomed to exact dates. The imprecise or even unknown dates of prehistoric stone and metal archeological finds hint at what an archeological world would look like without tree-ring dating. In the Royal Museum for Central Africa in Tervuren, Belgium, where I worked while pursuing my PhD, the dates of even many of the wooden artifacts are given as "unknown." The majority of the museum's masks, sculptures, neck rests, and stools were made in Central Africa, where to this day no reliable living-tree chronology is available with which to date even twentieth-century pieces. If it weren't for Douglass and his team's efforts to find HH-39 and to bridge the gap, the chronology of Ancestral Puebloan cultural development in the American Southwest and many other regions might still be equally unknown.

By joining the dated living-tree chronology with the floating chronology, Douglass also extended the tree-ring record for the Four Corners region back in time by more than 500 years, to 700 CE. This continuous, precisely dated record provided him with more than 1,200 years of data to study sunspot cycles and variations in climate. In the following years, he focused his efforts on extending this record even further back in time, and by 1934 he had succeeded in covering almost the entire Christian calendar (11–1934 CE). In 1937, building on more than three decades of dendrochronological achievements in archeology and climatology, Douglass established the LTRR at UA, the first department to be fully dedicated to the study of tree rings. UA housed the new laboratory under the bleachers on the west side of the football stadium and promised Douglass that this housing would be temporary until a more appropriate location could be found. When I arrived in Tucson in 2011, the LTRR was still under the bleachers of the football stadium. If you visit the stadium to watch a UA football home game, you will still find a door with my name on it on the west side. It was not until 2013 that UA made good on its more than 75-year-old promise and moved the LTRR to a brand new building.

Dendrochronology as a field of scientific expertise has grown a lot since its humble origins in southern Arizona in the 1930s. In addition to locking the American Southwest's prehistory in time, dendrochronology has been used as a precise dating tool in numerous archeological and art-historical projects, to check the precision of radiocarbon dating, to study the climate over the past 2,000-plus years, to put twentieth- and twenty-first-century droughts and pluvials* in a historical context, to study past earthquakes, volcanoes, wildfires, and other natural hazards, and in forest-history research. This smorgasbord of applications has only been made possible because since the LTRR inauguration more and more dendrochronology labs have been established worldwide. There are more than 100 data-producing tree-ring labs around the world, many of which put more than one experienced tree-ring researcher to work. The LTRR, for instance, is currently home to more than 15 faculty members, all of whom specialize in tree-ring research, in addition to about 50 administrative, technical, and curatorial staff, graduate students, postdocs, and outreach docents. Other large tree-ring labs are spread throughout North and South America (e.g., at Columbia University in New York City, the University of Mendoza in Argentina, and the University of Victoria in Canada), Europe (e.g., at the Swiss Federal Institute for Forest, Snow and Landscape Research, WSL, at Swansea University in Wales, and at Wageningen University in the

---

*A pluvial is a multi-year period marked by high rainfall.

Netherlands), Russia (e.g., at the Siberian Federal University in Krasnoyarsk), Asia (e.g., at the Chinese Academy of Sciences in Beijing), and Australasia (e.g., at the University of Auckland in New Zealand).

The florescence of dendrochronology across the world has resulted in the proliferation of exactly dated tree-ring chronologies. Luckily, dendrochronologists tend to have a strong collaborative spirit. We understand that the whole is greater than the sum of its parts and are generally happy to share our hard-earned tree-ring data with one another and with the broader science community on the International Tree-Ring Data Bank,* a publicly accessible internet database hosted by the NOAA (National Oceanic and Atmospheric Administration) paleoclimate program. By compiling the findings of almost a century of dendrochronology, the Tree-Ring Data Bank holds data in reserve from more than 4,000 sites. Tree-ring chronologies cover much of the Earth's land surface, especially in the Northern Hemisphere, and extend from hundreds to thousands of years back in time.

A century of dendrochronology, however, has also shown us many of the challenges and limitations of the field. As we have learned what works and what does not work in dendrochronology, it has become increasingly clear that when Douglass moved to the American Southwest, the conditions were, perhaps counterintuitively, just right for the establishment of tree-ring research. For instance, ponderosa pines—the species most often used by Douglass and by scientists to this day—have very distinct rings, are abundant and widespread throughout the Southwest, and are relatively long-lived. Trees 350 to 400 years old are fairly easy to find in the Southwest, and the oldest known ponderosa pine tree in Arizona was 742 years old when it was sampled in 1984. As with most other trees in the Southwest, the pine's growth in any given year is largely determined by how much water it receives, and the trees are very good recorders of year-to-year variability in precipitation. In wet years, with lots of snow and rainfall, the trees grow well and form wide rings. In dry years, they suffer and form narrow rings. The amount of precipitation in the Southwest varies strongly from year to year, and the sequence of alternating wet and dry years creates a sequence of narrow and wide rings that is shared

* At www.ncdc.noaa.gov/paleo/treering.html.

What Is Crossdating?

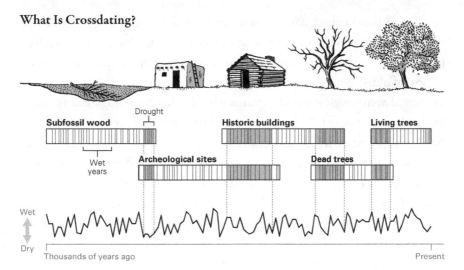

Figure 2 Alternating wet and dry years create a sequence of narrow and wide rings shared by trees from the same region, allowing us to match tree-ring patterns in live and dead trees, historic buildings and artifacts, and even charcoal and subfossil wood in a process called crossdating.

among southwestern trees, because they all experience the same wet and dry years. This shared tree-ring pattern is like a long string of Morse code (wide-wide-narrow-narrow-narrow-wide) that is very distinct and recognizable in all southwestern wood. It allows the tree-ring patterns in live and dead trees, in historic buildings and archeological wood, and even in charcoal and subfossil wood* to be matched to one another in a process called *crossdating* (fig. 2). For instance, we know that the entire Southwest was parched in the year 1580. We know this from the evidence of an extremely narrow ring in most of the trees and wood in the region and even in California's giant sequoias. The year 1580 can therefore be used as a *pointer year*, or benchmark, for dating wood of unknown age and for verifying the tree-ring sequence in living trees.

In addition to the abundance, longevity, and drought sensitivity of ponderosa pines, the Southwest has the additional advantage that Ancestral Puebloans used pinewood extensively in their constructions, resulting in its preservation

---

*Wood that is not yet fossilized, such as wood found on the bottom of lakes.

in ruins. This archeological wood served as a vehicle to link Douglass's tree-ring work in living trees to the field of southwestern archeology, which received growing attention as it accrued one new discovery after the next. The combination of old, drought-sensitive trees with distinct rings and abundant, well-preserved archeological wood explains why dendrochronology was founded in the Arizonan desert. If the hotspot for late nineteenth-century astronomy had been located where the forests were more biodiverse than in the American Southwest, the tree rings less distinct and less uniquely influenced by drought, and the prehistoric ruins sparser and less well preserved, the field of dendrochronology might have emerged through an entirely different sequence of events. And I might never have transplanted my life from the Swiss Alps to the Sonoran Desert.

▨ These dendrochronological benefits of the Southwest are largely lacking in the American Midwest, where Florence Hawley Ellis, the first woman dendrochronologist, pioneered tree-ring research. Hawley took Douglass's inaugural dendrochronology course at UA in 1930. After a few years of dendrochronological work at Chaco Canyon and completing her PhD, she took up a professorship in 1934 at the University of New Mexico, where she stayed until her retirement in 1971. Hawley was the undisputed trailblazer of dendrochronology east of the Mississippi River. She spent years collecting archeological wood samples from the burial and ceremonial mounds of the Mississippian culture (ca. 900–1450 CE) and sampling more than a thousand living trees in the Midwest in order to develop a living-tree reference chronology. As often happens when trailblazing, Hawley and her team faced a range of technical challenges. They were working in a broad and climatically diverse region with new, deciduous tree species whose rings were often indistinct and difficult to crossdate. In addition to this, a lot of the eastern and midwestern forests had been cut during eighteenth- and nineteenth-century European immigration, resulting in a shortage of old-growth forests and therefore old trees to develop long chronologies. There was also a shortage of archeological wood material because, unlike in the Southwest, organic material like wood is not well preserved in the damp mound sediment of many midwestern archeological sites.

Working in the late 1930s and the early 1940s, Hawley and her team also ran into challenges that were less technical in nature but were instead related to the culture of the era. During the Second World War, a member of Hawley's team was accused of being a German spy by landowners in western Kentucky. The landowners became suspicious when they came upon him coring trees on their land and subsequently found a German textbook while searching his car. Another member of Hawley's team then staged a coup in an attempt to get his hands on her research position, leveraging the precarious stature of women in science to undermine her. In an attempt to justify and hide his sexist behavior, he wrote in a letter to Hawley's boss: "Lewis wants and needs a *man* for the job. Please, . . . keep all this under your hat. No one but a few know of the situation that Hawley was involved in here last year and it would only hurt her for it to get out so I am asking you to depend on me that everything is OK."[*] I am pretty confident that an email exists somewhere featuring similar wording about me. A lot has changed, and luckily improved, in the 80-plus years since Hawley became the first female dendrochronologist, but unfortunately, some challenges for women in science have remained the same.

---

[*] Roy Lasseter to Sid Stallings, 5 March 1936, emphasis in original, from *Time, trees, and prehistory: Tree-ring dating and the development of North American archaeology, 1914–1950*, by Stephen E. Nash (Salt Lake City: University of Utah Press, 1999), 227.

Two
I Count the Rings Down in Africa

One Thanksgiving weekend a few years ago, I was driving from Tucson to Santa Fe, New Mexico, with two of my friends, both fellow UA scientists, for a much-needed break from midsemester madness. To entertain ourselves during the long drive, we played a game inspired by the twitter trend of the moment, #sciencesongs, in which you replace real song lyrics with references to your research. I was having a hard time finding appropriate lyrics for Rachel and Dave, who study plant-soil microbiology and the terrestrial carbon cycle, respectively. But in a tribute to dendrochronology's accessibility, Rachel triumphed with an adaptation of Toto's 1982 hit "Africa." The lyrical mastery of "I count the rings down in Africa"* is now a running joke between us, even if it demonstrates that my close friends think I stand around counting rings for a living.

Rachel's cleverness referenced the fact that, despite my mother's misgivings, I collected my first tree-ring samples in northwestern Tanzania in July 1998 during an epic field campaign with Kristof Haneca, a fellow master's student. Tanzania at the time was a blank spot on the world tree-ring map, and we were brazen enough to think that we could change that. We aimed to collect tree samples in Tanzania's dry seasonal forest to investigate whether local trees formed rings, and if so, whether they could be used to study East African climate.

Neither Kristof nor I had ever traveled outside Europe. We had no idea which tree species to sample, how to find trees to sample or even how to sample them, how to persuade local authorities to work with us, or how to ship potential samples back to Belgium. But what we lacked in expertise, we made up for in enthusiasm. We packed up a few tree-ring borers, a couple of bow saws, and a GPS and boarded a plane to Dar es Salaam, Tanzania's largest city.

*The original lyrics are "I bless the rain down in Africa."

One thing we did know was that we had one week to reach the city of Kigoma, 750 miles west of Dar es Salaam on the banks of Lake Tanganyika, if we were to catch up with our sole contact in Africa, a scientist from the Africa Museum in Belgium, who would help us settle in. Our trip started with a 20-hour flight delay, followed by the news that a railroad bridge had been wiped out by heavy rains and that the first stretch of our 36-hour train ride to Kigoma had been replaced by a three-day bus ride. This was when I learned the first rule of fieldwork: always make a plan but always be prepared to change it. When we finally made it onto the train to Kigoma, we were so relieved that we simply dropped our backpacks onto the floor of the train and turned to enjoy the view out the window. When we turned back around half a minute later, one of our backpacks was gone, and with it our GPS and tree borers. We still had our passports and our traveler's checks, but we had swiftly learned another lesson: not to leave our gear out of sight even for a second. After a week-long journey, we arrived in Kigoma on the same day that our only contact left the city. We were quite literally left to our own devices, which didn't amount to much without our GPS and borers.

Kigoma is a small city on the northeastern banks of Lake Tanganyika, about 40 miles south of the border with Burundi. In between Kigoma and Burundi lies Gombe Stream National Park, where Jane Goodall conducted her research on chimpanzees. Kigoma's main artery leads from the lakeside train station where we arrived to the next town, Ujiji,* an old slave-trade hub five miles down the road. The main road is lined with small restaurants and shops, as well as one hotel, the market, the bank, and the bus station. For a city of its size and remoteness, Kigoma hosted a surprising number of nongovernmental organizations (NGOs) and was home to many foreigners, who were called *mzungu* by the locals. Kristof and I stood out among the mzungus because we were on a shoestring budget and walked everywhere, including the five miles to Ujiji and then the four miles to the Kigoma airport, where the meteorological station was located. We must have appeared to personify the literal translation of mzungu: "aimless wanderer."

But Kristof and I did have an aim for being in Kigoma: we were there to collect tree-ring samples. Most tree-ring sampling involves an increment borer, a

---

* Ujiji is the place where Henry Stanley met David Livingstone in November 1871, introducing himself with the words" Dr. Livingstone, I presume?"

Figure 3 A tree-ring sample is extracted from a plains cottonwood (*Populus deltoides* ssp. *monilifera*) on the Yellowstone River floodplain in Montana using an increment borer and then stored for transport in a white paper straw. Photo by Derek Schook.

specialized, hollow borer used by dendrochronologists to extract cores from living trees or wood beams (fig. 3). Unlike chainsawing, coring allows us to extract the tree-ring information we need without harming or killing the trees or with only minimal damage to historic buildings. Unfortunately, Kristof and I had lost our increment borers on the train. Not wanting to cut down living trees for our research, we decided to look for trees that had recently been cut for charcoal production and to sample their logs or stumps. We were fortunate, dendrochronologically speaking: large stretches of forest around Kigoma were being clear-cut to produce charcoal, leaving an abundance of stumps for us to sample.

With the help of one of the Kigoma-based Jane Goodall NGOs, we selected a nearby recently cut site, where we met with the charcoal producers. They only spoke their mother tongue, one of the 124 languages spoken in Tanzania, which made communicating what we wanted, and especially why we wanted it, difficult. We managed to explain things with hand gestures and pantomime, making use of the hacksaw for demonstration. The charcoal loggers who made a living by cutting down and chopping up trees, however, were more comfortable and efficient with axes and machetes than with our saw. At the end of that day, we returned to Kigoma with two backpacks filled with roughly cut stem cross sections, or *cookies*.

We spent our last week in Kigoma collecting meteorological data, which involved transcribing by hand 70 years' worth of monthly temperature and rainfall data, and arranging the shipment to Belgium of the 30-something cookies we had collected. When we dropped off two big postal bags full of cookies at the train station, where they were to be put on the train to Dar es Salaam and from there shipped to Belgium, a large man picked up both bags as if they contained feathers rather than wood and threw them in a dark corner. In that moment, we did not think that the fruits of our weeks of hard labor would ever make it out of the Kigoma train station, let alone to Belgium.

Six months later, our samples still had not made it to Belgium. Kristof and I had started to panic and were working on Plan B for thesis topics when one day, miraculously, two postal bags full of cookies showed up on the steps of the Africa Museum. Only a few months away from graduating, we had very limited time to prepare, measure, crossdate, and analyze the tree rings and then write up our theses. I am not sure how, but we both managed to graduate a year after our Kigoma adventure. Our first field experience left a lasting impression on both of us and motivated us to continue, to experience more. We both went on to do PhDs in dendrochronology.*

⬛ Our Kigoma tree-ring work was pioneering and exploratory. We were the first to collect tree-ring samples in Tanzania and to investigate whether dendrochronology was possible in the Tanzanian woodland. If it was, our work could ultimately lead to *climate reconstruction*, in which tree-ring data are used to look at the climate of the past, the period before meteorological measurements. For climate reconstruction, the oldest, most climate-sensitive trees in the region would have to be targeted. In our exploratory work, Kristof and I did not have the luxury of being selective, of looking for the oldest possible trees. We collected as many samples as our limited budget and limited time allowed, without killing any trees in the process. We were lucky: the trees we had sampled showed distinct tree rings, and we managed to develop a 38-year-long, crossdated tree-ring chronology.

---

*Kristof is now a successful dendroarcheologist at the Flanders Heritage Agency in Brussels, Belgium.

Thirty-eight years' worth of data, however, does not provide much in terms of climate reconstruction. But such exploratory work, which examines which trees have crossdateable annual rings that reflect climate and its variability in new regimes, does form an important base for *dendroclimate* research. Precise, crossdated measurements of tree-ring width are essential for these studies, in which we need a data point and thus a ring-width measurement for each year in the past to guide us. With the groundwork covered, we can aim at reconstructing climate as far back in time as possible, by focusing our sampling strategy on old trees and dead wood on the ground.

When we sample old trees, we prefer increment borers over machetes or chain saws because they don't kill the trees or even harm them. Trees grow from the outside in. The most recent ring is just inside the bark, and the oldest, first-formed ring is at the center. It is the delicate layer between the bark and the wood, the *cambium* layer, that is responsible for a tree's woody growth. New wood cells are formed by the cambium and are deposited on the outside of a previously formed (older) layer of wood cells. Of the entire trunk of a tree, this wafer-thin cambium layer just inside the bark is the only part that is actually alive. Everything else—the wood and the bark—is dead material that functions primarily to provide stability and protection and to transport water and nutrients from the roots to the leaves and vice versa. Water transport occurs only in the outer area of the stem, the sapwood—not in its core—and is only marginally impacted by tree-ring coring. Likewise, tree-ring coring extracts only a very small area of the living cambium layer, typically only 0.2 inches in diameter, about the size of a chopstick, and the tree barely registers it.

Coring living trees with an increment borer can be physically demanding, especially when the trees belong to species with dense wood, such as oaks. We drill the borer into the tree manually, by twisting its handle, and once the wood core is removed, we remove the borer by twisting it in the opposite direction. This requires quite a bit of upper-body strength, especially if you're coring dozens of trees a day, and this often comes as a surprise to dendro newbies: experienced tree-ring corers are accustomed to the practice and don't always outwardly show the appropriate amount of exertion when demonstrating tree-ring coring. Without fail, at least one student in any group will complain about a defunct borer on his or her first attempt at coring, not realizing that it is the strength put in that is deficient, not the borer.

The physical exertion of coring, however, is rewarded with instant gratification once you extract the core from the tree. Unless your borer gets stuck or you hit a tree with rot in its center, you will pull out a core on which you can immediately see the rings. The knowledge that these rings will turn into the hundreds or thousands of data points that you'll use for your science is invigorating, and all without injuring the tree. If you are fortunate, the rings you see on the core will be narrow and manifold, by which you will know that you have cored an old tree. If you are even more fortunate, your borer will have gone straight through the center of the tree and you will have *hit the pith*, which means that you will have extracted the oldest ring that the tree has to offer.*

An experienced team of dendrochronologists can easily collect from 100 to 200 tree cores a day, depending on the size of the team, the size of the trees, the difficulty of the terrain, and so forth. Keeping tabs on who in the team hits the most piths motivates team members to try to extract the most cores and to do it right. After a good day of team coring, the cores can quickly pile up, and the resulting bundle containing hundreds of cores is a sight for sore arms. This is important, because after a long day of dendro-fieldwork you have a tangible result. Not a sheet of paper or a tablet full of numbers but actual, physical evidence of the hard work you put in. Eureka moments are rare in science. Most aspects of a scientist's job (writing papers, proposals, dissertations, books) involve slow, incremental steps. Given the general lack of instant gratification in science, that little jolt of palpability after coring a tree is invaluable.

■ Different trees tell different stories. An understory tree that has been living in the shadow of a taller tree for most of its life will mostly complain about its neighbor and be less bothered by the climate. A tree growing in a meadow might complain primarily about the goats or deer eating away at its foliage. A tree in a Mediterranean forest might complain about the wildfires that make its life miserable every few years rather than a particularly gloomy spring. But like people, many trees do enjoy talking about the weather. Trees in the American Southwest grumble when there's a drought, manifesting their discontent through narrow rings. Their counterparts in the Swiss Alps or the boreal forest in Alaska, however, are more displeased by cold summers than by drought and

---

* At least at the height that you core.

will record lackluster summer temperatures, rather than a shortfall of rain, in their tree-ring patterns. These "complaints," which limit a tree's growth, are termed *limiting factors* in the tree-ring world.

Dendroclimatologists aim to reconstruct climate as reliably as possible and pay close attention to limiting factors when selecting trees and sites to sample. We select trees whose yearly growth is primarily determined (or limited) by year-to-year variations in the climate and by little else. We choose overstory trees in sparse rather than dense forests because these trees don't suffer much from the complicating factors that come with competition from neighboring trees. We also prefer remote trees and forests that are hard to reach and that have been largely left alone by humans, because we ourselves can act as limiting factors. For instance, when we aimed to reconstruct rainfall variability in southwestern Virginia, we stayed clear of trees directly along the Appalachian Trail, where hikers often remove tree branches for firewood. Losing branches can be a stressful factor for a tree and, like fire scars or grazing damage, can eclipse climate as its dominant limiting factor.

To capture the strongest climate signal, we go to areas with harsh climatic conditions, where the amount a tree grows in a year, and thus the width of its ring, is limited by the weather conditions in that year. If we want to study past droughts, we will sample trees in dry regions, where trees suffer from lack of precipitation, rather than cold. On the other hand, if we want to reconstruct past temperatures, we will select sites in cold environments, where summer temperatures are not always optimal for the trees. This is why temperature reconstructions are largely based on trees from sites at high latitudes (Siberia, northern Canada, Scandinavia) or high elevations (the European Alps). Drought reconstructions, on the other hand, primarily originate in Mediterranean and monsoonal regions, where precipitation is limited and shows a clear seasonality.

In search of old trees growing in remote locations under harsh climatic conditions, dendroclimatological fieldwork takes tree-ring scientists to the most austere and beautiful landscapes. It often involves long hikes on steep slopes to reach remote areas with breathtaking views. Most tree-ring scientists, when asked about their favorite part of the job, will answer, "The field." It is what attracted many of us to this line of research in the first place, and it is what keeps us coming back for more.

Three
Adonis, Methuselah, and Prometheus

Old trees have common characteristics that are recognizable in the field, or even in a photograph. That makes it easier for us. Rather than coring every tree in a forest in the hope of finding the old ones, we can target the most conspicuous specimens, saving ourselves, and the trees, a lot of coring. When looking at old trees, a connoisseur will notice traits that many have in common: an untapered, columnar stem with few but heavy branches; large, exposed roots; and a dead top. Some old trees grow in spirals, and their bark grows in strips. Like humans, old trees (more than 250 years old) look different from middle-aged trees (50–250 years), which look different from young trees (less than 50 years). Also like humans, trees grow in height only when they are young but continue to grow in girth as they age. How tall a tree will grow and how long it will take for a tree to reach its full height is largely determined by its genes. A coastal redwood (*Sequoia sempervirens*) will grow taller than the cherry tree (*Prunus cerasus*) in your backyard. In fact, you can probably fit eight cherry trees in Hyperion, a coastal redwood that at nearly 400 feet is the world's tallest known living tree. Tree height is further influenced by the soil in which the tree grows, as well as by competition from other trees. Yet, a tree will only grow to a height that falls within the range of its species.

Because trees reach their maximum height and thereafter grow only in girth, young trees that are still getting taller have a tapered look. The top of their stem is only a few years old and has not had time to put on many rings and much girth, whereas the bottom part of the stem is older and has had more time to add rings and grow thicker. Once height growth has stopped in an older tree, then the upper part of the stem will start to catch up, its girth increasing year after year, and the stem will gradually take on a more columnar, rather than tapered, look. In addition to this, the very top of the tree will die off, and the crown will flatten out on top, not unlike a bonsai tree. The tree's limbs also continue to thicken; branches and roots of old trees often are quite sizable. The tree might lose lower branches that are overshadowed by its higher ones and do not contribute much to photosynthesis and growth. Centuries of erosion may have exposed the roots of old trees, so that often they are no longer underground. Some old conifers grow like a helix: instead of straight up vertically, their new wood cells grow at an angle, resulting in a spiral grain. *Spiral growth* has been attributed to genetics as well as to a variety of stress factors (such as asymmetrical crowns, wind, and slope) and is a hindrance in commercial wood production. Because it is impossible to follow the helix of spiral grain with an increment borer, spiral growth also significantly hinders dendrochronology. And yet, the hunt to find the oldest, most difficult trees to core has definite lures. The thrill of the chase can eclipse the challenges dendrochronologists face in the course of fieldwork, and sometimes even seems to usher us toward success.

In July 2015, we organized a 10-day field campaign in the Pindos Mountains, in northern Greece, in search of old, high-elevation trees. For me, the Pindos field campaign was part of my National Science Foundation–funded project aimed at using tree rings to reconstruct jet-stream patterns. I had been looking into the connections between tree rings and broader climate systems, and I knew my investigation would require sampling some of the oldest trees in Europe. To that end, I got in touch with Paul Krusic, who had recently sampled some very old trees on the Balkan Peninsula. Paul is an accomplished tree-ring scientist at Cambridge University, and seasoned when it comes to fieldwork. A mellow MacGyver type, he is easygoing and ingenious, with chainsaw skills beyond compare and a passion for Land Rovers. A few years earlier, Paul had stumbled upon a photograph of some very distinctive-looking Bosnian pines (*Pinus heldreichii*) on Mount Smolikas, the highest peak in the Pindos Mountains. In his mind there was no question that the pines were old, as they displayed all the stereotypical features of old growth; on top of that, they were growing in a steep, rocky landscape. Paul knew he had to go and have a look at the trees for himself.

When Paul returned to Cambridge after his first Smolikas field trip, he counted, to his astonishment, more than 900 rings on one of the tree-ring samples he had taken. Regrettably, the longest borer Paul had taken on his exploratory trip had been considerably shorter than he needed to reach the pith of that more than 900-year-old tree. There are many practical reasons why we only take the longest borers, those 40 inches in length, into the field under special circumstances: (1) they are expensive and scarce; (2) they are heavy and add considerably to the weight of a field pack; and (3) they are wider in diameter than regular borers (16–24 inches), which makes them even harder to bore into a tree, and so the risk of getting them stuck is higher. Therefore, to increase efficiency and reduce risks, we use the shortest possible borer needed to reach the pith of any given tree. Only on rare occasions is a yard-long borer required. Paul had just identified such an occasion! There were older rings still to be discovered deeper inside that ancient tree. The only question was, how many? To answer this question, Paul and I decided to join forces and go on a second sampling spree to Mount Smolikas.

We returned to Smolikas with a vengeance and armed to the teeth. Or more precisely, with a bigger team, longer borers, and most importantly a chain saw to sample dead wood. On his first Smolikas trip, Paul had noticed lots of sun-bleached dead wood littering the landscape. These were the relic trunks of dismembered tree ancestors that had died hundreds of years earlier and had been lying around since, slowly eroding. It was exciting to imagine that this material could potentially be used to extend the Smolikas tree-ring chronology even further back in time.

Paul drove down to Greece from his home in Stockholm in his Land Rover with his son Jonas, a precocious 12-year-old Viking who, as a notetaker, sample labeler, and occasional corer was a huge help in the field. Our team was completed by two scientists from the University of Mainz in Germany: my former adviser Jan Esper, a 6-foot-4 charismatic German with a peculiar sense of humor, and Claudia Hartl, who, despite her petite appearance, is a formidable force in the field: she is strong, organized, tireless, and never complains.

We were lodged in the tiny town of Samarina, the highest village in Greece, at the foot of the mountain. On the first morning, the steep 2- to 3-hour hike from our breakfast place in town up to the treeline site on Mount Smolikas seemed insurmountable. But on the mountain ridge the sight of ancient trees in an utterly barren environment awaited us. As the days wore on, our pile of samples grew larger, and the morning's hike became easier. The weather at tree line on Mount Smolikas, at 6,500 feet, was sunny and mild. As a bonus, the forest was mercifully free of mosquitoes, and we managed to avoid encounters with any of the feral dogs, wolves, or bears that were said to roam the mountain. Also, we were in Greece, where the food is delicious, the people are friendly, and the wine is abundant. At the end of each long day of coring and cutting, we would hike back down, replace our boots with flip-flops, and enjoy a marvelous meal before literally stumbling into bed.

We cut more than 50 cookies and cored more than 100 trees during those 10 halcyon days, including the ancient tree Paul had originally discovered. Yet, it took a third field campaign in 2016, with an even bigger team, and finally a 40-inch borer before Adonis—named after the Greek god of beauty and desire—revealed its true age: *more than 1,075 years.* That fateful day in Greece, we extracted a core measuring nearly 3 feet (!) from Adonis and still did not quite hit the pith. Discovering Europe's oldest known living tree was a fitting capstone for the epic Mount Smolikas field campaigns, no matter how many trips it took to achieve.

When we first discovered Adonis on Mount Smolikas and spread the word that this 1,075-year-old senior was the oldest known living tree in Europe, we received a lot of backlash mainly from two fronts: the "clonal tree" community and the "heritage tree" community. Both communities claim that much older trees can be found and reliably dated in Europe. The right party in this debate, in my opinion, depends on our definitions of *tree* and *dating.* Clonal trees, which propagate and disperse asexually through root suckers, are genetically identical and share a common root structure. Such root structures can be more than 10,000 years old, as determined by radiocarbon dating, but the individual stems rarely reach more than a few hundred years in age. For instance, Pando, a clonal colony of more than 40,000 stems originating from a single quaking aspen (*Populus tremuloides*) in Utah, has a root system estimated to be 80,000 years old, but its individual stems rarely reach 130 years. Depending on whether or not one includes clonal root systems in the definition of *tree*, Europe's oldest living tree could be Sweden's Old Tjikko—a 9,550-year-old clonal Norway spruce (*Picea abies*) named after its discoverer's dog— or Adonis, the individual tree we dendrochronologically dated at just over 1,000 years old.

I first came across the heritage tree dispute when I lived in Switzerland, many years before I was involved in the discovery of Adonis. My friend Frank had invited me to his house for dinner to meet his new girlfriend, an art teacher and hobby photographer. All went well until Frank mentioned their upcoming vacation to the British countryside to visit the ancient yew (*Taxus baccata*) trees, with which his girlfriend had been enamored for years. Frank's girlfriend claimed that some of these yews were more than 3,000 years old, and she cited entire websites dedicated to the ancient yews to back up her claim. The atmosphere grew tense when I started arguing that if there were 3,000-year-old trees in Great Britain, then I—as a professional who dated trees for a living—would certainly know about them. The feisty dendrochronologist in me cast doubts on her claim, while she vehemently insisted that she wasn't just making things up. Needless to say, the evening ended earlier than anticipated.

When I got home, I browsed the websites Frank's girlfriend had referred to and soon discovered the issue at the heart of our heated discussion. Heritage trees are typically large, old, solitary trees with unique cultural or historical value. The yews found in churchyards across the British Isles are one good example. Others include the ancient olive trees (*Olea europaea*) found across the Mediterranean Basin and the Hundred Horse Chestnut (*Castanea sativa*), growing on the slope of Mount Etna in Sicily. The cultural importance of heritage trees is illustrated by the Sicilian poet Giuseppe Borrello (1820–1894), who wrote about the origin of the Hundred Horse Chestnut's name:

A chestnut tree
Was so large
That its branches formed an umbrella
Under which refuge was sought from the rain
From thunder bolts and flashes of lightning
By Queen Giuvanna
With a hundred knights
When on her way to Mount Etna
Was taken by surprise by a fierce storm.
From then on so was it named
This tree nestled in a valley and its courses
The great chestnut tree of one hundred horses.

There is no doubt that heritage trees are old, but their exact age is generally elusive. Many trees that are monumental in girth are fragmented and have lost the oldest core part of their stem to rot. This makes dendrochronological or even radiocarbon dating impossible, and instead tree ages are estimated based on size or extrapolated from presumed growth rates. Since growth rates differ significantly between trees and even within trees—young trees grow much faster than old trees—these estimates can be inaccurate. As a result, the longevity of heritage trees is often overestimated and subject to intense dispute. For instance, British yews can easily reach an age of 600 or even 800 years, but that is a far cry from the 4,000 to 5,000 years claimed for the Llangernyw Yew in Clwyd, North Wales, or the 3,000 to 9,000 years attributed to the Fortingall Yew in Perthshire, Scotland. In theory it is possible that some European heritage trees are more than a thousand years old, but I have yet to find convincing evidence. I have learned over the years, however, to be more compassionate when it comes to other people's passion projects, and I think we can safely conclude that Adonis is the oldest, *dendrochronologically dated* tree in Europe.

There is one additional factor to take into account when considering the age of heritage trees: location. The world's oldest known dendrochronologically dated trees grow in remote, barren environments with strong environmental limiting factors. It is rare that trees growing in mild environments, such as the Welsh countryside, or in places with a long human cultural history of wood use, such as Sicily, reach a very old age. There is a reason for this. Trees in harsh environments are severely limited in their growth, and they grow slowly—very slowly. Adonis, for instance, on average grows less than a tenth of an inch (about 1.5 mm) in diameter per year. This slow growth results in very narrow rings and relatively dense wood. The dense wood of slow-growing trees, which is often resinous in conifers, makes them more resistant to invasions by insects, fungi, and bacteria and therefore less prone to decay. Fast-growing trees, on the other hand, such as eucalypts (*Eucalyptus* spp.) and cottonwoods (*Populus* spp.), are programmed to make optimal use of spring weather by producing a lot of wood quickly. Such *pioneer species*—so called because they are often the first to colonize open spaces—thus produce a lot of *earlywood*, the light wood formed in spring. The large vessels of earlywood are optimal for bringing water from the roots to the emerging leaves, allowing trees to grow quickly, up to an inch in diameter per year for eastern cottonwoods (*Populus deltoides*). The hastily formed wood of these pioneer trees is light, soft, and weak, and it easily succumbs to pests. Pioneer trees honor their name: they work hard, play hard, and die young. These trailblazers follow Neil Young's credo that it's better to burn out than to fade away.

It is the faders among trees that persevere through time. Bristlecone pines (*Pinus longaeva*) are perseverance personified. These stunted and twisted trees quite literally fade out as they age. More and more of their rings go missing, and they focus their growth energy on fewer and fewer branches. More and more of the root-stem-branch connections in aging bristlecone pines die off, and the oldest trees hang on to life by a few narrow strips of living bark that connect only a few individual roots to a few individual branches. Such *strip barking* allows the trees to conserve their energy so that they do not succumb to stem or limb damage caused by fire, lightning, or extreme weather. The Great Basin, in the western US, is home to a handful of bristlecone pines that are more than 4,000 years old, and the oldest living bristlecone pine is dendrochronologically dated to be over 5,000 years old, making this species the longest-lived on earth.

Bristlecone pines grow in isolated groves on the dolomite outcrops of dry and exposed mountainsides, an environment in which hardly any other plants can survive, and where the pines themselves are few and far between. The landscape is sparse and the trees gnarled, with scant foliage and wood with a glazed appearance from centuries of erosion by wind and water. Viewed against the blue sky in this dry mountain environment, the bristlecone pines look austere and otherworldly, and they speak to the imagination of artists and writers. The oldest bristlecone grove can be visited in the Ancient Bristlecone Pine Forest, a protected area in the White Mountains in eastern California that was established in 1958, a year after the LTRR's Edmund Schulman discovered its ancient inhabitants. As you can imagine, crossdating or even sampling the bristlecone pines, with their distorted stems, narrow strips of living tissue, deep crevasses of dead wood, and missing rings, is a complex process. Schulman, who had been working with Douglass at the LTRR since the 1930s, was well prepared for this challenge; he had spent most of his summers in the mountain ranges of the American West looking for old trees. His search culminated in the discovery of Methuselah, at 4,789 years the then oldest known living tree, which he named after the legendarily long-lived biblical patriarch. Dating from 2833 BCE, Methuselah still graces the Ancient Bristlecone Pine Forest, but in order to avoid vandalism, its exact location is not disclosed to visitors (not even visiting dendrochronologists).

Schulman died at age 49, only a year after discovering Methuselah. In fact, a number of bristlecone pine researchers have died relatively young, including the LTRR researcher Val Lamarche (1937–1988) and a 32-year-old Forest Service employee who suffered a fatal heart attack on his way down from sampling the pines up in the mountains. This morbid coincidence spawned a long-standing urban (or rather, sylvan) legend that the wood of the bristlecone pine carries a curse; it is whispered in some circles that whoever works with these, the oldest trees, will die young. Happily, this myth has been disproved by some of my venerable colleagues, but another classic tale of bristlecone pine dendrochronology has turned out to be a nightmare in truth.

Seven years after Schulman discovered Methuselah, an even older bristlecone pine tree was discovered on Wheeler Peak, in eastern Nevada, in what is now Great Basin National Park. At 4,862 years, Prometheus, as the tree was named by local mountaineers, was 73 years older than Methuselah, displacing it as the oldest known living tree of its time. According to Darwin Lambert, one of the mountaineers, fewer than fifty people saw Prometheus while it was still alive. The tragic fact is that its age was not discovered until it had been cut down. That's right, in 1964 the oldest known living tree in the world was cut down. To count its rings.

Don Currey, who at the time was a graduate student in geography at the University of North Carolina, was interested in dating and analyzing the bristlecone pines in eastern Nevada to advance his study of Holocene-age glaciers in the American Southwest.* The first tree he stumbled upon when he arrived at Wheeler Peak with his borer was Prometheus. There are differing accounts of what happened next. Some say his borer was too short or got stuck; others say that he did not understand how to core such a large and distorted tree or that he preferred a full cross section for his studies. Regardless of the motivation, he asked for—and received—permission from the US Forest Service to fell Prometheus. Later that night in his hotel room, Currey counted 4,862

---

*The Holocene is the current geological epoch, which started approximately 11,650 years ago.

rings on the cross section he had cut from Prometheus and realized with horror that he had just killed the oldest known living tree on earth.

When news of Currey's unforgivable blunder came out, public outrage ensued. In a 1968 *Audubon* magazine article entitled "Martyr for a Species," Lambert called Currey a murderer. Understandably, Currey has switched research topics and has dedicated the rest of his scientific career to the study of salt flats. In a rare media appearance on PBS's *NOVA* in 2001, Currey described the moment he realized Prometheus's age: "You've got to think, 'I've got to have done something wrong. I better recount. I better recount again. I better look really carefully with higher magnification.'" But no matter how many times he recounted, Currey always came out with more rings than anyone had ever counted on a tree before. It would take almost another half century for a living tree older than Prometheus to be found. Finally, in 2012, LTRR researcher Tom Harlan sampled a bristlecone pine, this time using a borer and leaving the giant standing, and determined it to be 5,062 years old, dating its innermost ring to 3050 BCE. The tree's identity and exact location are a secret.

The high and dry environment of the Great Basin not only empowers bristlecone pine trees to grow for millennia, it also helps to preserve the wood of dead trees after they expire. Not many other plants grow in this barren landscape, and the scarce ground cover and litter mean that wildfires are rare. The bare limestone rock on which the trees grow is not hospitable for wood-decaying fungi and insects, so the resinous wood can remain on the landscape for thousands of years after a tree has died. For a few summers in the early 2000s, LTRR researchers took volunteers up to the White Mountains to find and sample remnant bristlecone wood as part of a project sponsored by an anonymous bristlecone aficionado. Some of the remnant wood turned out to have originated from trees that died more than 8,000 years ago and to have been lying around on the mountains since then. Crossdating the dead wood to the living-tree chronology extended the bristlecone pine chronology back to 6827 BCE. The 8,800-plus-year-long chronology is extensive enough to study climatic changes in western North America and their long-term influences on forest ecosystems over millennia. The unique precision of such millennia-long, continuous tree-ring chronologies is also invaluable in calibrating other dating methods, such as radiocarbon dating, and other paleoclimate records, such as ice cores.

Given the stories of Methuselah and Prometheus, it isn't surprising that out of all regions in North America, the West hosts the most long-lived tree species. What is surprising is that not all of these long-lived trees are stumpy and gnarly like the bristlecone pines. Some long-lived trees of the West, such as the giant sequoias and the coastal redwoods in California, are majestic titans that lack the stereotypical visual characteristics of old trees. The giant sequoia was one of the first species Douglass sampled in 1915, in Kings River Canyon in the Sierra Nevada mountains. The sequoia groves there had been heavily logged in the preceding decades, so Douglass collected large wood sections from the remaining stumps, which were sometimes more than 30 feet in diameter. The oldest stump Douglass sampled was 3,220 years old. Despite the ongoing heavy logging, there are likely giant sequoias of equal age or older still standing in the Sierra Nevada mountains, but even a 40-inch borer—such as Paul Krusic used for Adonis—will not suffice to determine the exact age of these living behemoths. The other giants of the American West, the coastal redwoods, can live for more than 2,200 years and are joined in the 2,000-plus league by at least three other species in the region: the western juniper (*Juniperus occidentalis*), the Rocky Mountain bristlecone pine (*Pinus aristata*, a different species from the previously discussed bristlecone pine), and the foxtail pine (*Pinus balfouriana*).

Western North America's geography certainly contributes to its abundance of long-lived trees, since many old trees prefer dry mountain slopes, but proximity to humans may also impact our perspective on what constitutes abundance. The American Southwest is the birthplace of dendrochronology, boasting a long history and a high density of tree-ring research projects. In addition to this, the American West's history of large-scale deforestation did not start until the 1800s, making it short compared with that of other regions. The vast majority of old-growth forests in Europe, for instance, were cut in Roman times, so it is no surprise that no trees remain that date prior to then. Finally, the well-established infrastructure in western North America also helps dendrochronologists find old trees. Even the most remote areas in the western US are easier to reach than their counterparts in, for instance, Siberia or Tanzania. Trust me, I've tried.

The oldest living tree in North America is more than 5,000 years old, whereas the oldest tree in Europe is barely more than 1,000 years old. On almost all other continents we can find old trees that fall roughly within this age range. A Huon Pine (*Lagarostrobus franklinii*) in Tasmania is the oldest living tree in Oceania. Ed Cook, known as the godfather of dendrochronology, and his team at Columbia University's Lamont-Doherty Tree Ring Lab sampled it in 1991, when it was just under 2,000 years old. Its innermost sampled ring was dated to 2 BCE. In Africa, an Atlas cedar (*Cedrus atlantica*) in the Atlas Mountains of Morocco was dated to be 1,025 years old. It is surpassed in age by a baobab (*Adansonia digitata*) in Namibia that is estimated to be 1,275 years old, but there is a potential error of 50 years on the baobab's age because the tree was studied through radiocarbon dating of its pith, not by tree-ring dating. Radiocarbon dating is not as precise as tree-ring dating, but baobab wood does not show distinct growth rings, so this method provided the best possible estimate for the tree's age. Asia's oldest known living tree (1,437 years when sampled in 1990) is a juniper (*Juniperus* spp.) growing at the upper timberline in the Karakorum in northern Pakistan.

I have little doubt that on each of these continents there are older living trees that we are not yet aware of, but it does appear that the oldest trees in absolute age are found in the Americas. In South America, El Gran Abuelo, an alerce (*Fitzroya cupressoides*) growing in the Alerce Costero National Park in Chile, was tree-ring dated in 1993 to be 3,622 years old. In eastern North America, bald cypress (*Taxodium distichum*) are the largest and longest-lived trees. The oldest known living bald cypress was at least 2,624 years old when Dave Stahle, from the University of Arkansas, sampled it in 2018 along the Black River in North Carolina. Sampling these giants that grow in swamps and can easily reach 150 feet in height and 12 feet in diameter is not for the fainthearted. To get above the sometimes 10-foot-high swollen "knees," or *buttresses*, that surround the base of the trunk, Dave ascended the trees with climbing spikes, which penetrate the bark but do not harm the tree. "I've probably climbed a thousand cypress trees that way," Dave once told me. He uses a system of ropes and lanyards to secure himself to core the trees above the buttress. "It is very difficult to get into position and to tie yourself up so that you can push and pull an increment borer effectively," Dave explained, "but the one saving grace about *Taxodium* is that it's like coring butter. Once you get your corer in, the actual coring isn't so hard."

Despite the wet conditions in which bald cypresses grow, they are good recorders of change in the level of swamp water. When water levels are high, the trees enjoy good water quality with high levels of dissolved oxygen and nutrients. When water levels are low and water quality is poor, this is reflected in their narrow tree rings. Unfortunately, the massive bald cypress trees produce large quantities of attractive pale brown to reddish wood that is resistant to decay and very desirable for construction. In the late nineteenth and early twentieth centuries, commercial cypress exploitation boomed in the southeastern US, leading to excessive logging. Today, only three reasonably sized tracts of old-growth cypress remain: one sanctuary each in South Carolina and southern Florida and one tract on private property in Arkansas.

There are therefore few options left for dendrochronologists who want to find old bald cypress wood. One option is *sinker wood*. The exploitation of bald cypress in the late nineteenth century involved river-floating the hand-cut logs from their original forest stands downstream to mills. Many of these logs from long-lived, straight-grained trees never made it to the mills because they got stuck or sank along the way. This created caches of century-old logs on the bottoms of many rivers in the southeastern US. With no living old-growth trees left to cut, such sinker wood is unrivaled in quality and can command thousands of dollars. Despite the dangers related to extracting sinker wood from river banks and bottoms, including alligators and water moccasins, companies are popping up throughout the Southeast that specialize in harvesting sinker wood. The right dendrochronologist could potentially persuade such companies to slice off a sinker cookie for dendro-dating before turning the rest of the log into highly coveted furniture. An even more ambitious option, one that calls for perseverance, is hinted at in the bald cypress's nickname, "wood everlasting." All along the Atlantic coastal plain of the Southeast, non-petrified, subfossil bald cypress stumps and logs have been recovered from buried deposits. One such buried bald cypress deposit, the Walker Interglacial Swamp, was exposed just four blocks from the White House in downtown Washington, DC. The Walker Swamp stumps, whose roots are still attached, have remained in their upright growth position for ca. 130,000 years. However, with the longest bald cypress chronology currently covering ca. 2,600 years and the longest continuous tree-ring chronology in the world adding another 10,000 years to that time period, we have a long way to go before we can bridge the gap with the Walker Swamp stumps.

Four
And the Tree Was Happy

We know that trees can live for millennia, and we have a pretty good idea of how to detect those ancient trees in a landscape and how to sample them (hint: not with a chain saw). But how do we go about extracting dendrochronological information from these trees? How do we turn their longevity into information we can use to date valuable violins or reconstruct past temperatures?

In their 1977 book *Climates of Hunger* Reid Bryson and Thomas Murray write, "Nature does not make mistakes in the records she leaves. We sometimes do not understand them properly; that is the source of the difficulty." Trees remember. They record history and they don't lie. But to correctly interpret the stories trees tell us, we need to read their ring patterns with the attention they deserve. This requires a little talent for pattern recognition, a lot of training and concentration, and a proper understanding of what makes trees tick. Luckily for dendrochronologists, trees are relatively straightforward organisms. They originated in a geological epoch long before humans arose, when life was simple. Trees have far fewer moving or redundant parts than humans—no tailbones and no male nipples. To find the wealth of information trees have to share, we must simply learn how to look.

Once painstakingly gathered tree-ring samples make their way from the field to the lab, the dendrochronologist's first step is to glue the cores into wooden core mounts, which stabilizes them and ensures an even core surface to look at through a microscope. Next comes sanding: it is difficult to differentiate narrow rings from one another and to take precise ring-width measurements on rough, unsanded samples. The woodworkers among you will appreciate that we sanded the cookies we collected in Tanzania incrementally from 80 to 1,200 grit* and then polished

---

* Grit number refers to the quantity of abrasive particles that can fit through a square-inch filter. Coarse sandpaper has a low grit number, whereas fine and superfine sandpapers have high grit numbers.

them. This superfine sanding might have been a bit of overkill, as most samples are sanded to 400–800 grit. But for our Tanzanian samples, we needed extra clarity to detect the boundaries of the rings, which were often just a few cells wide.

When you look at finely sanded wood through a microscope (and I highly recommend that you do), you can see individual wood cells and even the details of their cell walls (fig. 4). Trees are extremely well oiled carbon-capturing machines with an elegant simplicity that is reflected in their physiology and the anatomy of their wood. Each cell has a specific and indispensable function. In conifers, individual wood cells queue up in straight lines like Roman soldiers, arranged for optimal strength and functionality. In broadleaf trees, which evolved later, the wood cells form more intricate and eye-catching patterns that are unique for each tree species, so that a trained wood anatomist can identify a tree species by looking at its wood alone.

Trees grow more vigorously in spring, when they are well rested after a good winter sleep, than they do in fall, when they are getting ready for a winter-long dormancy. The wood formed in spring, *earlywood*, reflects the tree's energetic spring growth. Conifers form large earlywood cells with thin walls,* whereas broadleaf trees form earlywood vessels that are specifically designed for water transport. The earlywood of both conifer and broadleaf trees is thus optimized to transport water and nutrients from the roots to the growing canopy in spring. Later on in the growing season, structural support and carbon storage become more important to trees than water transport, so *latewood* cells, formed in late summer and fall, are smaller with thicker cell walls. Some broadleaf tree species (most dramatically, oak) form much larger vessels in the earlywood than in the latewood. This sequence of large earlywood vessels and smaller latewood vessels makes for clear annual tree rings and for beautiful, *ring-porous* wood. Earlywood vessels in oak are often so large that you can see them with the naked eye, for instance, on the end-grain surface—typically the short end—of a solid oak table.

This sequence of an earlywood growth spurt in spring, a transition to latewood growth in fall, and then a halt of growth in winter dormancy happens every year in the life of a typical tree in a temperate forest. The abrupt transition from last year's small-cell latewood to this year's large-cell earlywood creates a distinct boundary between rings that separates one year's growth from the next

---

*Wood cells in conifers are called *tracheids*.

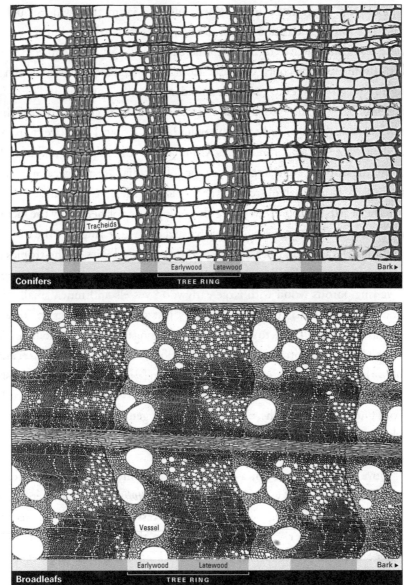

(A) Conifers — Tracheids — Earlywood — Latewood — Bark ▶ — TREE RING

(B) Broadleafs — Vessel — Earlywood — Latewood — Bark ▶ — TREE RING

Figure 4A Conifers form tracheids that are square-shaped and form neat lines as the tree grows. The earlywood cells are large with thin walls; latewood cells, which form in late summer and fall, are smaller with thicker walls.

Figure 4B In broadleaf trees, the wood cells form more intricate patterns unique to each tree species. The circular areas are vessels, cells designed to transport water. The sequence of large earlywood vessels and smaller latewood vessels in some species, such as oak, creates clear annual rings and a beautiful wood grain.

and allows us to see and count tree rings and to measure their widths. Such distinct tree-ring boundaries are often absent in trees that grow in climates without clear seasonality, such as the tropics, where the day length and temperature vary little throughout the course of a year. Tropical climates are also wet enough and warm enough for trees to grow year-round. As a result, many tropical trees don't necessarily feel the need for a yearly dormancy or to form clearly distinguishable earlywood, latewood, and tree-ring boundaries. Tropical trees are therefore a challenge for dendrochronologists. Compared with temperate and boreal regions, the tropics (including, for instance, Tanzania) are a huge, almost blank spot on the world map of tree-ring chronologies. Here be dragons . . .

The dearth of tropical tree-ring chronologies is not helped by the fact that in the tropics there are very few conifer species, whose rings are the easiest to read. But like all good rules, the no-tree-ring-chronologies-in-the-tropics rule has exceptions. Teak (*Tectona grandis*) trees in Southeast Asia, for instance, have ring-porous wood with large earlywood vessels and much smaller latewood vessels and form beautiful tree rings, just as temperate oaks do. Teak was used as early as the 1930s to develop centuries-long tree-ring chronologies. Motivated by entirely different seasonal conditions, Arapari (*Macrolobium acaciifolium*) trees growing in Amazonian floodplain (*varzea*) forests in Brazil show distinct rings, as their growth is interrupted every year by flooding that lasts from four to eight months, causing anoxic (oxygenless) conditions in the soil, which trigger a growth dormancy.

▨ I love trees, but I am not a tree-hugger. The only times I refer to trees as sentient beings are (*a*) when I am reading *The Giving Tree* to my nephew and (*b*) when I am explaining crossdating, the process of comparing one tree's ring pattern with another's. Trees are happy when they have plenty of food and water and no one is competing with or attacking them. In a happy year a tree will grow a lot and form a wide ring. In a less happy year—there might be a drought, or it might have been cold, or a hurricane might have blown off its leaves and branches—the tree will not have much energy to grow, and it will form a narrow ring. The happiness of trees is thus strongly influenced by climate. Trees suffer not only from seasonal affective disorder—they hibernate during the dark season—but also from annual affective disorder: they get depressed in years when the climate is bad. Whether those "bad climate" years for trees are defined by cold or dry conditions depends on the region. In semi-arid regions such as the American Southwest, trees get depressed and grow narrow rings in dry years. In alpine and arctic regions, on the other hand, cold years, rather than dry years, will result in narrow rings. But within a given region, bad climate years, whether they are dry or cold, will affect most trees in the same way and will leave a narrow ring across the board.

In the American Southwest, for instance, dry years result in narrow rings in most trees, whereas those same trees will form wide rings in years of rainfall plenty. The climatic succession of wet (happy) and dry (unhappy) years is recorded as a recognizable pattern of alternating wide and narrow rings in the trees, the Morse code I referred to in chapter 1. This long string of code is the pattern we try to match between samples when we crossdate. The crossdating process involves visual or statistical pattern matching or, most often, a combination of both. We can measure the width of all individual rings of the samples we aim to crossdate and then compare the measured tree-ring series to find the best statistical match. For precise measurements, we use a moveable, digital measurement stage that allows us to measure and record individual ring widths with a mouse click (fig. 5). Measuring ring widths, however, is time-consuming and labor-intensive. Experienced scientists working on applications for which dating (unlike, e.g., climate reconstruction) is the primary aim, such as dendro-archeology, can actually circumvent measurements and statistical crossdating, relying instead on their own memory and pattern-recognition skills to cross-date tree-ring patterns visually, without digital aids.

The visual crossdating process often starts with what Douglass referred to as *tree-ring signatures*, which are distinctive sequences of consecutive narrow and wide rings, snippets of Morse code, with obvious pointer years that stand out in a signature pattern. When he started to date southwestern archeology samples, Douglass would first check undated samples for the presence of the tree-ring signatures that he knew and recognized, such as the 611 (narrow)–615 (narrow)–620 CE (narrow) signature. If this 610s signature stood out in a sample, it would give him a place to start his crossdating and to anchor the undated sample in time (fig. 6). To put Douglass's tree-ring signature in a human-history perspective, it was in the 610s that Muhammad received divine revelations and started preaching the Quran in Mecca, thus introducing the religion of Islam to the world. A trained dendrochronologist will often know by heart the tree-ring signa-

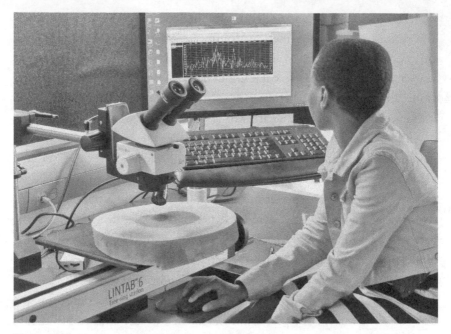

Figure 5 Dendrochronologist Zakia Hassan Khamisi measures the ring widths on a *Brachystegia spiciformis* cookie from Tanzania using a digital measurement stage connected to a desktop computer, which also allows her to simultaneously plot the measurements. Photo courtesy of Laboratory of Tree-Ring Research, University of Arizona.

tures and the centuries-long sequence of narrow and wide rings common to most trees in a given region. Just by looking at it she can match the pattern she sees in an undated sample under the microscope to the pattern that lives in her head.

◼ When I was working on samples from the Californian Sierra Nevada, the tree-ring signature for the late eighteenth and early nineteenth centuries looked like this:

> 1783: narrow
> 1792: wide
> 1795: narrow
> 1796: very narrow
> 1809: wide
> 1822: narrow
> 1829: very narrow

The rings between the above pointer years were typically of an unremarkable width. The most prominent pointer year of the signature was 1796, the last year of George Washington's presidency, which must have been an enormously dry year in the Sierra Nevada, because its ring was very narrow in almost every sample I looked at. When analyzing a new, undated sample, the first thing I would do is look for the narrowest ring I could find. Knowing that the ring for 1796 was very narrow on almost every sample, the plan was then to see whether the narrowest ring in this particular sample could possibly be the one for 1796. Most of the samples we collected in the Sierra Nevada derived from stumps of trees that were cut during the mining boom in California, sometime between 1850 and 1900. I thus knew to look for the 1796 narrow ring about 50 to 100 rings (or years) in from the edge—the outermost, most recent ring—of the sample.

The next step was to see whether the ring prior to the narrowest ring, which I assumed to be the 1796 ring, was also narrow, like the 1795 ring in the tree-ring signature. If this was the case, then I would count back another three rings and see whether the 1792 ring was wide. Then count back another nine years and see whether the 1783 ring was narrow, and on back through time. I'd also look forward in time—13 rings in the outward direction to see whether 1809 was wide, 1822 narrow, and so forth. It would soon become apparent whether the sample's pattern of wide and narrow rings matched the pattern in the tree-ring signature and thus whether I was right in my assumption that the very narrow ring I had detected was for 1796. If so, great: I could just continue to match up all the pointer years back to the innermost ring of the tree, which frequently dated back to the 1400s or even earlier. Then I could match up all the pointer years forward into the nineteenth century, determine the year of the outermost ring, and consider the sample dated. The 1783–1829 Sierra Nevada tree-ring signature was not sufficient on its own to reliably crossdate a sample, but it gave me a place to start. If the first narrow ring I had selected did not fit the Sierra Nevada tree-ring signature, then I would start the whole exercise again, looking for a different narrow ring that could potentially be for 1796. And then a different one and another one, and yet another one until I found a match with the tree-ring signature. At times, especially in the beginning, this was a long and frustrating process, but it allowed us to date about 90 percent of the 2,000 samples we collected in the Sierra Nevada.

Tree-Ring Signatures

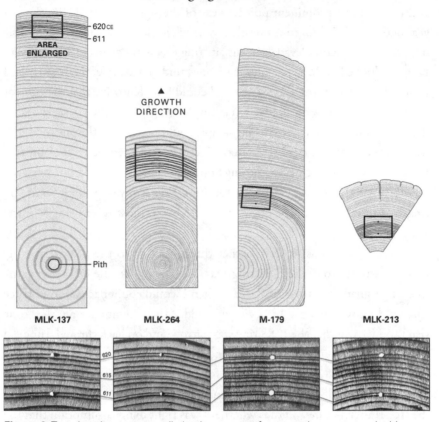

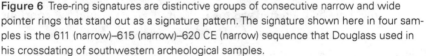

Figure 6 Tree-ring signatures are distinctive groups of consecutive narrow and wide pointer rings that stand out as a signature pattern. The signature shown here in four samples is the 611 (narrow)–615 (narrow)–620 CE (narrow) sequence that Douglass used in his crossdating of southwestern archeological samples.

The way to memorize the tree-ring signatures for a given region, such as the 1783–1829 Sierra Nevada tree-ring signature, is to look at very many samples from that region and then to look at more samples, and then even more. No two tree-ring series are identical. Even two samples from the same tree will not be identical! But if samples come from the same area, they will have some pointer years in common, and these pointer years will become more and more obvious as you study more and more samples. When you investigate what sometimes seems like an endless number of samples, the years when many trees have abnormally narrow or wide rings begin to stand out, and before you know it, you have the sequence of pointer years for the region in your head, or at least on paper. At the end of the two-year Sierra Nevada project, after crossdating almost 2,000 samples, I definitely knew how to find that 1796 narrow ring. I could walk into the lab, grab a sample out of the stack of undated samples, and put my finger on the 1796 ring before I had even looked at it under a microscope.

We apply the same principle of sample replication when we build a reference tree-ring chronology, against which we can crossdate new, undated samples. A reference chronology is anchored in the present time, because it is based on living trees with a known date for the most recent ring,* and provides the key to the regional Morse code. The more samples the reference chronology includes, the better it reflects the common pattern across samples, because the idiosyncrasies of individual samples, the randomness in non-pointer years, will have been averaged out. Typically, when we develop a reference chronology for a new area, we sample at least 20 trees and often many more to have enough samples to represent the common tree-ring pattern.

Just as we need a sufficient number of samples for reliable crossdating, we also need a sufficient number of years covered by each sample. The longer-lived the sampled tree or wooden beam, the more rings are available for crossdating. The longer a tree has lived before being cut or sampled, the more years with anomalous climate and growth it will have experienced, and the more pointer years there are for us to match against a reference chronology. The more matching pointer years, the less room for mistakes or uncertainty about where the sample matches the reference chronology and ultimately about its dating. You can think of it like solving a jigsaw puzzle: the more jagged and multifaceted the edge of a piece, the fewer places it will fit. In reality, there is only one correct solution to the puzzle, only one correct match between sample and reference chronology. Each tree has grown only once, and so its tree-ring pattern fits at only one point in the chronology timeline. You can try to fit that jagged piece in a spot where it does not belong, but you'll have to cram it in with force; it will still stick out of the puzzle, and you will only be fooling yourself. Tree rings don't lie.

---

*The year the tree was cored.

The excitement of dating a piece of wood just by looking at its ring pattern is also akin to that of solving a puzzle. It requires a similar level of concentration: it is unlikely that your crossdating will be efficient or even successful if you haven't slept much or if (just for instance) you are forced to listen to loud jam-band music coming from your labmate's Phish-inspired Pandora station. The crossdating process is the true heart of dendrochronology—this is why it is a science and not just counting rings. Once your brain is trained, crossdating can be an utterly satisfying activity, even if the road to get there is rocky and the learning curve is steep.

Another obstacle to deal with when crossdating, unrelated to the dangers of tie-dyed labmates, is the occasional occurrence of *missing rings* and *false rings*. Some trees that are not very stress resistant just give up during extremely dry years. Rather than forming an excruciatingly narrow ring, the tree just skips forming a ring altogether. It is as if the tree's heart skips a beat, and the result is a missing ring. Such missing rings are more frequent in drier environments and in older trees. For example, the year 1580 CE was so dry throughout the Southwest and California that only a minority of trees actually formed a ring in that year. To a trained eye, most missing rings are fairly straightforward to detect through crossdating. If you know what a tree-ring pattern is supposed to look like, then a hiatus in that pattern will jump out at you. You will notice if the tree-ring pattern you are looking at through a microscope has skipped a beat compared with the tree-ring pattern in your head.

The opposite can happen as well: a tree can occasionally form more than one ring in a single year. Such false rings are common in summer monsoon climates, where pre-monsoon droughts in late spring may trick the tree into thinking that the year has somehow fastforwarded to the fall. In this case, the tree starts forming its denser latewood prematurely, before the summer monsoons hit, then it realizes its mistake and reverts to forming the larger earlywood cells. When the actual fall arrives, the tree then forms a second band of latewood cells in the same year. Under a microscope, the premature, pre-monsoon, false latewood band is easy to distinguish from the true latewood. The transition between a false latewood band and the ensuing monsoonal earlywood is gradual, whereas the transition between real latewood and the following year's earlywood is razor-sharp. False rings are much more common in some tree species (for instance, juniper) than in others, and like missing rings they can be detected through crossdating. If missing and false rings go undetected, however, they can really wreak havoc in the timeline and, therefore, the accuracy of a tree-ring chronology.

■ As a rule, crossdating is easiest and most successful in regions where a single dominant climatic limiting factor impacts most trees in the region in the same way. If most trees in a region share a dominant limiting factor, then their Morse code will match and crossdating will work well. My first attempt at crossdating in our Sierra Nevada project went smoothly because California experiences very severe droughts (for instance, the recent 2012–16 drought or the 1796 drought) that leave their mark in most trees. Ponderosa pines in the American Southwest also experience severe and frequent droughts, which is why it was possible for Douglass to develop and apply the concept of crossdating.

Crossdating, however, can also work in mysterious ways. The reasons behind its effectiveness are not as obvious everywhere as they are in the case of southwestern pines. The world's longest continuous tree-ring chronology up to now is based on subfossil European and sessile oaks (*Quercus robur* and *Quercus petraea*) found in gravel pits in Germany. The subfossil tree trunks are remnants of forests that once grew along large, German rivers (the Rhine, the Main, the Danube) but over time were undercut by erosion. When trees topple over in water or peat, their wood is preserved in the anoxic environment, where oxygen and thus wood-decaying organisms—which need oxygen to breathe—are missing. The oaks and pines that once populated these forests, the vast majority of which were less than 300 years old when they died, have long since been buried by sediments, gifting us with remnants that have now been crossdated back more than 10,000 years. *The German oak chronology comprises 6,775 samples and dates back continuously from the present to 8480 BCE, thus covering more than 10,500 years without a single gap.* It crossdates with even older subfossil Scots pine (*Pinus sylvestris*) samples found in the same region, which extended the chronology by approximately another 2,000 years.

To develop a chronology of this length on a continent where the oldest living tree is barely over 1,000 years old is quite a feat in and of itself, but it becomes even more inconceivable if we expand the scope of our story to include research conducted in the British Isles by the University of Belfast dendrochronologist Mike Baillie. Mike developed a 7,272-year-long tree-ring chronology based on ancient (European and sessile) oak wood preserved in Irish peat bogs. As it turns out, the German and Irish oak chronologies crossdate with each other over the full 7,000-plus rings of overlap. Yet we do not fully understand why. We don't know yet what synchronizes the German and Irish oaks' heartbeats, what their common limiting factor is. While summers in Germany and Ireland are wet and warm enough for oaks to be happy most of the time, not producing many pointer rings, their summer climates are not that similar: dry and wet summers in Germany do not necessarily co-occur with dry and wet summers in Ireland. Nevertheless, over at least the past 7,000 years the oaks have shown clear and recognizable corresponding patterns of wide and narrow rings. European oak is one of the most studied species in dendrochronology, yet after 40-plus years of research, the driver of this large-scale growth synchronicity has yet to be discovered.

■ The German and Irish subfossil oak wood in river deposits, lakes, and peat bogs is only in the first stage of the petrification process, which ultimately takes millions of years. A surprising number of stems and stumps of ancient trees have been preserved over time as *petrified wood*, for instance, in the Petrified Forest National Park in eastern Arizona. Petrified wood is fossil wood in which all organic material has been replaced by mineral deposits—often quartz or calcium—and in which the original structure of the wood has been maintained. In order for wood to turn into stone like that, it needs to be buried under sand or silt sediment or volcanic ash so that oxygen cannot reach it and it remains preserved. Over time, water laden with minerals flows through the sediment covering the wood and deposits some of these minerals in the wood cells. The mineral deposits form a stone mold inside the wood cells, and when the organic cell walls decay, an internal, three-dimensional cast of the cells and the wood as a whole remains. Along with the wood-cell structure, tree rings are also preserved in their stone casts and, amazingly, are often as clearly visible as if they had formed the previous year rather than millions of years ago. These rings in petrified wood can be measured and even crossdated just like present-day tree rings. But given that petrified wood is millions of years old, while the longest present-day tree-ring chronology only spans the most recent 12,000 years, tree rings from the two archives will never be connected across time.

Yet the study of petrified tree rings (*paleodendrochronology*) can tell us a lot about the forest, the climate, and the environment in which the ancient trees grew. For instance, petrified wood has been found in Antarctica, where the current glacial landscape is too cold and dry to support trees, but this has not always been the case. During the late Permian and Triassic geological periods (from roughly 255 million to 200 million years ago), as well as more recently during the Cretaceous and the Paleogene periods (from about 145 million to about 23 million years ago) the climate on the Antarctic Peninsula was sufficiently warm and wet to harbor diverse plant communities, which included conifer forests and even broadleaf trees in the later periods. During the earlier periods, Antarctica was joined with other current-day Southern Hemisphere continents in the ancient supercontinent of Gondwana. Due to plate tectonics, Gondwana began to gradually break up some 180 million years ago, but South America, the most recently formed continent, did not separate from Antarctica until approximately 30 million years ago. As a result of this long shared geological history, relatives of the long extinct Antarctic plant families, such as Antarctic beech (*Nothofagus antarctica*) and the Podocarpus and Araucaria genera, can be found in southern South America, southernmost Africa, and Oceania.

The Antarctic petrified wood is derived from tree species that no longer exist but once grew on a supercontinent that itself no longer exists. However, the petrified wood remains from 145 million to 23 million years ago that do still exist are irrefutable evidence of a warmer South Pole and the absence of major ice sheets at these times. The fact that anatomically distinct tree rings can be detected in the fossils indicates that the Antarctic climate had well-defined seasons and that the warmer periods were not caused by a drastically reduced tilt of the Earth's axis, as has previously been proposed. Instead, climate models show that the high Antarctic (and Arctic) temperatures during the Cretaceous and the Paleogene can only be explained by increased carbon-dioxide levels in the atmosphere, which would not only have raised temperatures at the poles but also enhanced tree growth. The explanation for the wide tree rings found in the petrified wood could therefore be that the Antarctic forests were growing in a former greenhouse climate, a natural analogue to the contemporary climatic conditions we are now creating by emitting vast amounts of greenhouse gases. Whereas paleodendrochronologists' investigations into these ancient petrified tree rings may help us gain new insights into Earth's climate shifts millions of years ago, tree-ring science has more frequently been used to study much more recent climate and history, and what humans have been up to since we arrived on the scene.

Five

The Stone Age, the Plague, and Shipwrecks under the City

The epic German oak-pine chronology, spanning the past 12,650 years with annual precision, delivers a data point for each and every year all the way back to 10,641 BCE. Not only does it provide an exact, absolute date for each of its rings but its accuracy makes it an invaluable tool for calibrating other, less precise dating methods, such as radiocarbon dating.

Through radiocarbon dating (also called "carbon dating" or "carbon-14 dating") the age of archeological objects containing plant- or animal-derived material can be determined up to about 50,000 years ago. Radiocarbon is a radioactive carbon isotope that is created in the Earth's atmosphere through the impact of cosmic rays, which are highly energetic particles traveling through space at close to light speed. Radiocarbon undergoes radioactive decay with a half-life (the period of time it takes for the original amount of radiocarbon to be halved) of 5,730 years. It is incorporated in the tissue of living plants through photosynthesis and then in the tissue of animals that eat these plants. But once the plant or animal dies, its radiocarbon is no longer exchanged with its environment, and the amount of radiocarbon in its tissue slowly decreases as it decays. In the late 1940s—a couple of decades after Douglass established tree-ring dating—the chemist Willard Libby figured out that because we know the half-life of radiocarbon, we can calculate when a plant or animal died by measuring the amount of relict radiocarbon in its remains, such as a piece of wood or a bone fragment. Libby received the 1960 Nobel Prize in Chemistry for his discovery, which revolutionized the field of archeology. Compared with tree-ring dating, radiocarbon dating can date much older objects, but with less precision. Radiocarbon dates are typically presented as an age range that can span decades to centuries, whereas tree-ring dates are precise to the year.

The idea behind radiocarbon dating—that we can determine the age of an object because we can measure the amount of radiocarbon it contains and we know the half-life of radiocarbon—is valid only if the amount of radiocarbon in the atmosphere is constant. But we know that this is not the case. For instance, there has been a distinct decline in the amount of atmospheric radiocarbon since the late nineteenth century, when we started burning fossil fuels in vast quantities. Coal, oil, and gas are formed over millions of years from the remains of plants and animals. Because this million-year conversion process from living material to fossil fuel is much slower than the radioactive decay rate of radiocarbon, fossil fuels have lost almost all of their radiocarbon over time. By emitting large amounts of fossil fuel–derived carbon dioxide that contains close to no radiocarbon, we have substantially diluted the proportion of radiocarbon in the atmosphere. The reverse happened in the 1950s and 1960s, when we dramatically hiked up the amount of radiocarbon in the atmosphere by conducting nuclear tests. There is a spike in the atmospheric radiocarbon record in 1963, the final year of aboveground nuclear testing; in that year the amount of radiocarbon in the atmosphere was almost twice as high as prior to the nuclear tests.

Not only man-made but also natural fluctuations in the amount of atmospheric radiocarbon have occurred over time, and as a result radiocarbon-dated years do not directly equate to calendar years. The accuracy of the radiocarbon-dating technique therefore needs to be tested for different time periods in the past by comparing (calibrating) the radiocarbon age to the real age of objects for which we know the absolute, definite age. Tree-ring samples have lent themselves perfectly to this calibration exercise, because crossdating ensures that we know the real age of each ring. In any given year, trees add new wood material, with new atmospheric radiocarbon, only to their outermost ring. The radiocarbon content of all previous rings remains untouched, and these earlier rings reflect only the atmospheric radiocarbon of the year in which they were formed. Because of radiocarbon decay, which occurs in older rings even during a tree's lifetime, the amount of radiocarbon gradually decreases from a tree's outermost (most recent) ring to its innermost (oldest) ring. The radiocarbon content of the innermost ring of a 5,000-year-old bristlecone pine, for instance, will be only about half that of its outermost ring, according to radiocarbon's 5,730-year half-life. By measuring the radiocarbon content in the wood of individual tree rings, a radiocarbon calibration curve can be developed that relates a ring's radiocarbon content to its absolute age. This calibration curve can then be used to derive an estimate of any archeological object's calendar age from its radiocarbon content.* Libby himself started such calibration efforts in the 1960s, and the calibration curve is still actively updated today. The most recent version of the calibration curve uses radiocarbon measurements of tree-ring samples from the German oak-pine chronology to cover the most recent 13,900 years† and plant macrofossils preserved in thin-layered sediments of Lake Suigetsu in Japan for the older part of the curve (from 13,900 to 50,000 years before present).

The 12,000-plus-year-long German oak-pine chronology has not only helped us to fine-tune radiocarbon dating; it has also provided an absolute timeline for wooden archeological finds that recount the story of more than 7,000 years of wood use in European culture. Some of the earliest wooden settlements in Europe date back to the Neolithic (the new stone age, starting ca. 6000 BCE), when the practice of agriculture first spread across the continent. Many early Neolithic farming communities developed around water sources and made ample use of wood for construction because it was a widely available resource that required no sophisticated tools for processing. Neolithic communities constructed small residences on top of posts or piles in lake or bog wetlands that were easily defensible. Such *pile dwellings* were constructed throughout Europe and were inhabited from the late Neolithic until the end of the Bronze Age, around 500 BCE. The wooden posts that supported the dwellings were easily driven deep into the wet marshlands or lake sediments and have remained waterlogged and preserved through time. Following the discovery of such dwellings at Lake Zurich in 1854—at the same time as the rise of southwestern archeology in the US—excavations of pile dwellings proliferated throughout Europe and the British Isles. In the 1960s, two Swiss lake settlements were radiocarbon dated to ca. 3700 BCE, and the realization that these humble pile dwellings had been built more than 1,000 years before the Egyptian pyramids caused a stir in the archeological world.‡ The earliest dendro-

---

* As long as the object is less than 50,000 years old.

† The oak-pine chronology is supplemented by a floating chronology from 232 German and Swiss trees to extend the calibration to 13,900 years.

‡ The earliest known Egyptian pyramid is the Pyramid of Djoser, which was radiocarbon dated to 2630–2611 BCE.

chronologically dated pile dwelling, located in Lake Murten, in Switzerland, was constructed of oaks that were felled between 3867 and 3854 BCE.

The 7,000-plus-year-long British Isles bog oak chronology has also facilitated tree-ring dating of numerous prehistoric bog dwellings and crannogs,* as well as trackways that ancient British ancestors used to travel between settlements. The oldest known such pathway, the Sweet Track, was a raised wooden walkway, more than a mile long, consisting of a single line of oak planks held in place by wooden pegs. It crossed the Somerset Levels, a flat landscape extending over 270 square miles in western England, and was tree-ring dated to the winter and spring of 3807–3806 BCE, around the same time as the Lake Murten pile dwellings. The absolute dating of the Sweet Track, which likely was in use for no more than a decade before it was overrun by water and reeds, enabled the dating of a plethora of Neolithic artifacts (such as potsherds and flint and stone axes) found in the surrounding marshland.

The absolute earliest dendrochronologically dated wooden structure, however, predates the Lake Murten settlement and the Sweet Track by more than 1,300 years. In 2012, Willy Tegel, a German dendroarcheologist affiliated with the University of Freiburg, found four water wells in eastern Germany that were lined with wood walls. These wooden well constructions, which were underground and waterlogged, are the only tangible remains of the longhouses built by the first central European farmers. Of their aboveground construction, only the outline is left in the soil. The early settlers felled 300-year-old oak trees that were up to three feet in diameter and constructed well linings from the oak wood to prevent the wells from collapsing. To his surprise, Willy found that the water wells dated back to 5206–5098 BCE, not long after the first farmers immigrated into central Europe from the Balkans around 5500 BCE. The construction of such water wells included sophisticated corner joints and log constructions, which required advanced carpentry skills. As Willy put it in his 2012 publication: "The first farmers were also the first carpenters."

Oak well linings have also been preserved and tree-ring dated for later periods, including the Iron Age (ca. 800–100 BCE in central Europe) and the Roman period (ca. 100 BCE–500 CE). While still in Zurich, I collaborated

---

*A crannog is an artificial island built in a Scottish or Irish lake or river.

with Willy on a project in which we used the tree-ring data from his collection of Roman-era samples to reconstruct the climate in central Europe over the past 2,500 years. I would sometimes visit Willy in Germany to discuss the project. On one such occasion, I picked up a random piece of oak wood lying around his lab that was almost black and heavier than expected. Willy told me that he had crossdated this piece of a plank used in a Roman water well to 14 CE, and he persuaded me to take it as a souvenir. I was too concerned about damaging the 2,000-year-old chunk of wood to put it in my backpack, so I carried it in my arms. On the train ride home, I noticed a young boy staring at my log, so I explained to him that I was a tree-ring scientist and that the wood was 2,000 years old! He looked at me as if I had come from another planet.

■ There is a lot that we don't know about early European settlers. How large was their population? What language did they speak? How and why did they build megaliths such as Stonehenge? But thanks to dendrochronology, we do know the exact year and even the season when they cut oaks and pines to build pile dwellings, trackways, and water wells more than 6,000 years ago. And that is just the beginning. The amount of information that dendrochronology has contributed to our understanding of history picks up dramatically once above-ground construction wood is preserved. From medieval times onwards, timber from "dry" construction offers an abundance of puzzle pieces for dendrochronologists to play with. Tree-ring dating has become a key contributor to the study of castles and cathedrals, universities and city halls, as well as more modest historic buildings. From German Viking settlements to Venetian palazzi, England's Salisbury Cathedral, and the Hagia Sophia in Istanbul, dendrochronology has provided new insights into not only the structures but also the cultures of civilizations around the world.

Sometimes, dendrochronology can actually alter our perception of human history. The world's oldest wooden building, the Hōryū-ji Temple in Nara Prefecture in Japan, for instance, was generally believed to have been constructed at the beginning of the eighth century. Hōryū-ji is the oldest continuously active Buddhist sanctuary in Asia, and historical documents reveal that whereas the original temple was completed in 607 CE, it was destroyed by a catastrophic fire only 63 years later. Its beautiful pagoda, sought out by tens of thousands of visitors every year, was believed to have been reconstructed around 711 CE and has miraculously survived civil wars, earthquakes, and ty-phoons ever since. But in 2001 the dendrochronologist Takumi Mitsutani and his colleagues discovered that the Japanese cypress (*Chamaecyparis obtuse*) tree used for the central post of the pagoda was felled in 594 CE, more than a century earlier than historians had concluded. The gap between the dendrochronological date, 594, and the historical date, 711, is hard to explain, but it is possible that the current post was recycled from the original temple or that the cypress timber was stored unused for an extended period of time. Such a discovery could cause scholars of religion and history to reassess their timeline of the rise and spread of Buddhism in Japan.

In North America, in addition to extensive dendroarcheological work in the American Southwest and a traditional nineteenth-century Nuu-chah-nulth plank house on Vancouver Island, more than a thousand colonial structures have been tree-ring dated. These include historically significant buildings such as Independence Hall in Philadelphia (1753) but also more vernacular archi-tecture such as cabins, churches, homesteads, corrals, and trading posts. Such structures are often in fact one or two generations younger than originally doc-umented. For instance, one of the log houses at the Marble Springs Historic Site in Tennessee was once thought to have been the final home of John Sevier, Tennessee's first governor. Yet, when Jessica Slayton and her colleagues at the University of Tennessee examined the cores they extracted from the cabin's logs, they discovered that the cabin had not been built until more than twenty years after Sevier's death in 1815. This tendency for oral history to make things older than they actually are can be blamed on human nature: we like to see our heritage as venerable. Of course, there are economic drivers as well; older his-torical sites attract lucrative tourism to local communities.

▨ Historic buildings provide the majority of wood for dendroarcheology, but smaller wooden objects, such as doors, furniture, art-historical artifacts, and even medieval books with wooden covers can also be dated. The oldest den-drochronologically dated door in the British Isles, for instance, was dated to the eleventh century (1032–64). The almost 1,000-year-old door is still in ev-eryday use in Westminster Abbey in London, where it closes off an understairs cupboard. To examine it, Dan Miles and Martin Bridge, from the Oxford

Dendrochronology Laboratory, simply lifted it off its hinges and drilled into its edge with an increment borer.[*]

Other historic art objects, such as panel paintings, wooden sculptures, furniture, and musical instruments, such as Stradivari's Messiah, cannot be cored without visual damage, so ring widths are measured either directly on the piece or on photographs, scans, or imprints of its surface. It takes a dendrochronologist with stronger nerves and steadier hands than mine to remove an oak panel painted by a Flemish Primitive in the fifteenth century from its frame and prepare its edge with a scalpel or a Dremel to make the rings in the wood clearly visible. Sometimes laser or micro-abrasion equipment is used for this purpose, and this kind of preparation and sampling is obviously done in consultation with conservators and curators. The authenticity of works by most of the old masters who painted on oak panels from the fifteenth through the seventeenth century (Van Eyck, Memling, Bruegel the Elder, Bruegel the Younger, Rembrandt, Rubens, and so forth) has been verified using dendrochronology. If the most recent ring on an oak panel is dated to a time later than the date marked on the work, it means that the tree from which the panel was cut was still alive when the painting is claimed to have been created and the painting in question is likely a copy or a forgery. Dendrochronology may not be able to conclusively identify the artist, but if the tree it was ultimately painted on was still growing after the artist's death, it makes for a convincing refutation.

A fifteenth-century triptych altarpiece by the Flemish painter Rogier Van der Weyden (1400–1464) is a good example. There are two copies of the triptych. The right panel of the first copy is housed in the Metropolitan Museum of Art in New York, while its left and middle panels are in the Capilla Real in Granada, Spain. The second copy as a whole is housed in the Gemäldegalerie Dahlem in Berlin, Germany. On the basis of art-historical studies, it was always assumed that the New York–Granada triptych was the original, painted by Van Der Weyden himself, and that the Berlin version was a later copy. Tree-ring dating of both triptychs, however, dated the Berlin triptych to ca. 1421, early in Van der Weyden's life, whereas the New York–Granada triptych was

---

[*] To accurately align the drill bit of the borer with the boards of the door, the dendroarcheologists used a series of guides fitted to a jig clamped to the face of the door. They used compressed air to cool the drill bit and clear it of dust.

dated to 1482, almost two decades after his death. All along, the Gemälde-galerie in Berlin was displaying the Van der Weyden original, while admirers at the Met were looking at the work of a skilled unknown copyist.

Wooden artifacts from such a mix of places—doors from the British Isles, panel paintings from the Low Countries, Italian violins, to name just a few—cannot all be dated against one central European oak chronology. Rather, a network of chronologies is needed that densely covers the wide geographical area where artifacts come from and extends back far enough in time to date centuries-old objects. Luckily, as dendrochronology matures, tree-ring scientists have developed more and more chronologies and thus created the necessary network containing multiple tree species. This network is most dense in North America and Europe, where dendrochronology has been applied for many decades. Such a multi-species, geographically dense tree-ring network allows dendroarcheologists to investigate not only when an archeological or historical wooden object was made but also where the wood came from. The process of dendroprovenancing is based on the concept that a tree-ring series will statistically match more closely to chronologies from nearby sites than to remote chronologies. To provenance a wooden object, a dendroarcheologist can crossdate its tree-ring series to a network of chronologies from a wide geographical region and find the location with the strongest statistical match, which is then the suggested source region. This method is not foolproof and works better for some regions than for others, but it has seen remarkable success, for instance in nautical archeology.

Shipwrecks often yield timbers that are suitable for tree-ring research, but in most cases the wreck is found far from the place where the ship originally was constructed, which is often unknown. Dendroprovenancing has, for instance, shown that the *Karschau*, a medieval ship excavated in a fjord in northern Germany, was built from Danish trees felled in the 1140s. A shipwreck found in the 2010 excavation at the World Trade Center in Lower Manhattan in New York City was traced back to oak timbers felled in the Philadelphia region in 1773. It is likely that the ship was the product of a small shipyard and only had a short life span before it was sunk, deliberately or accidentally. In the 1790s, it became part of the landfill that was used to expand the buildable area of Lower Manhattan.

The shipwreck studies demonstrate how dendrochronology casts light on the long history of timber trading in the Americas and Europe. In western Europe, oak-beech forests were exploited intensively during the Middle Ages for the construction of castles, cathedrals, ships, and palaces, leaving old-growth, high-quality oak timber a scarce—and thus expensive—commodity. According to England's 1086 CE Domesday Book,[*] only 15 percent of the country was covered by woodland at the time. Large quantities of oak timber were therefore imported from elsewhere, particularly the Baltics, to support the western European medieval building frenzy. Wood was floated down the Baltic rivers to the Hanseatic ports, then loaded onto large seagoing ships, before being resold in the trade centers of western Europe. Even with so many middlemen involved, the cost of imported Baltic wood was typically only about a fifth of that of locally grown wood. Dendroprovenancing of medieval panel paintings and other western European historic works of art has demonstrated that the Baltic timber trade started as early as the late thirteenth century. After our joint adventure in Tanzania, Kristof Haneca started a PhD in dendroarcheology to study the wooden sculptures from Late Gothic altarpieces in northern Belgium. He found that early (fifteenth-century) altarpieces were carved out of wood originating from forests near Gdansk, one of the large Hanseatic harbor cities. Over time, however, the demand for oak wood from the Baltics became so high that easily accessible forests in the Gdansk region became overexploited. As a result, later (sixteenth-century) altarpieces were made out of wood originating from forests further inland. As the nineteenth-century German forester August Bernhardt described it, by late medieval times "timber famine was knocking on everybody's door."[†]

The historical timber trade is just one aspect of the complex history of wood use and building activity that can be documented by dendroarcheological research. By compiling the tree-harvest (or felling) dates in large tree-ring collections and lining up the number of trees felled per year, we can construct a timeline of tree harvesting and building activity (fig. 7). For the 2,500-year central European climate reconstruction that I worked on with Willy Tegel,

---

[*] A written record of the extent, value, ownership, and liabilities of land in England and parts of Wales, compiled by order of King William the Conqueror.

[†] August Bernhardt, *Geschichte der Waldeigentums, der Waldwirtschaft und Forstwissenschaft in Deutschland*, vol. 1 (Berlin, 1872), 220.

Building Activity in Europe
500BCE–2000CE

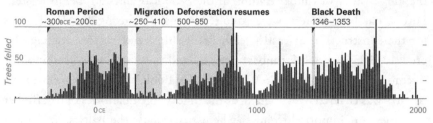

Figure 7 By compiling the harvest dates of almost 7,300 trees felled in Europe for construction purposes over the past 2,500 years, we can see historical phases of building activity. A short-lived building hiatus around 1350 CE can be attributed to the Black Death.

we compiled 7,284 tree-ring series from subfossil, archeological, historical, and recent oak wood from northeastern France and western Germany. By lining up the tree-harvest dates over two and a half millennia, we were able to identify historical phases of building activity. We found a high number of tree-harvest dates and extensive deforestation during the late Iron Age and the Roman period (ca. 300 BCE to 200 CE), reflecting an active building period. During the Migration Period, or *Völkerwanderung** (ca. 250–410 CE), a time when barbarian invasions contributed to the disintegration of the Roman Empire and to lasting political and social turmoil, tree harvesting and building activity diminished. Deforestation picked up again in sync with socioeconomic consolidation around 500–850. As figure 7 shows, there was also a short-lived building gap around 1350. This fourteenth-century building hiatus also is reflected in archeological tree-ring collections from Ireland and Greece. Such widespread, synchronous building gaps can only really be explained by one of two factors: a continent-wide socioeconomic collapse (such as happened during the Migration Period) or a pandemic. We know from other sources that the bubonic-plague pandemic (the Black Death) took Europe by storm in 1346–53 and culled 45–60 percent of its population. The demographic impact of the Black Death is thus reflected in a significant lull in the timeline of European building activity. An earlier (664–68 CE) outbreak of the plague in the British Isles can also be seen as a construction-timber hiatus in the Irish tree-

*Literally, "wandering of the peoples."

ring collection. Whereas such plague outbreaks were catastrophic for Europe's human population, they gave Europe's forests a breather from the relentless onslaught of deforestation. With more than 50 percent of Europe's population gone, the demand for energy and wood diminished, and forests were given a short-lived chance to recover and reclaim abandoned land.

Six
The Hockey Stick Poster Child

The Swiss Federal Institute for Forest, Snow, and Landscape Research—Wald, Schnee und Landschaft (WSL) in German—sits on top of a hill in Birmensdorf, just outside Zurich. Tree-ring research first became part of WSL's research mission in 1971, when Fritz Schweingruber started his work there. Fritz is a botanist, an archeologist, and one of the world's finest wood anatomists. Early in his research career he became fascinated with tree rings, and under Fritz's leadership the WSL dendrochronology group grew to be the European equivalent of the LTRR in Tucson. When I started working at the WSL in 2007, after my two-year stint at Penn State, Fritz had retired, but he still visited the dendro lab a few times a week, and his influence was omnipresent. Jan Esper, Fritz's protégé, who was later involved in the discovery of Adonis, had taken over the leadership of the lab, gathering a productive group of dendroclimate researchers around him. During my first summer at the WSL, Jan and I, along with two of these talented up-and-comers, David Frank and Ulf Büntgen,* traveled to the Spanish Pyrenees to collect tree-ring samples. The site we sampled was at almost 8,000 feet elevation, on the mountain slopes surrounding Lake Gerber in the Aigüestortes National Park. When I asked why we were sampling this particular site, the answer was simply that Fritz had recommended it. Apparently, while traveling through the Pyrenees, Fritz had spotted the site through binoculars from a road hundreds of meters below.

Fritz's renown for having an eye for old trees was well founded: as it turns out, the Lake Gerber tree-ring chronology, made up of living mountain pine (*Pinus uncinata*) trees and remnant wood, goes back more than a thousand

---

* David is now the director of the LTRR in Tucson; Ulf is a professor of geography at Cambridge University.

years. The Pyrenees field campaign was a swift introduction to the efficiency with which my WSL colleagues operated. We flew from Zurich to Barcelona, then the same night drove three and a half hours to the small town of Viella. At eight o'clock the next morning, we stopped for provisions and then hiked three hours straight up the mountain to the lake. Go-getters that they were, Ulf and David practically sprinted up the mountain, while Jan and I followed at a more reasonable pace. By the time we made it to the lake, it was noon and I figured we would stop for lunch before starting to core. But that was not how these guys rolled. The moment the first old tree came into sight, David started coring, with Jan quickly following suit. For the next few hours, David, Jan, and I cored one tree after another, while Ulf ran from one of us to the other collecting cores and then labeling and storing them. Finally, around three in the afternoon I flatly refused to carry on until we had lunch. Reluctantly, my three companions agreed to take a quick break, after which we continued coring relentlessly until the sun began to set and we had to rush back down the mountain to beat the dark. The exact same thing happened on each of the following days: we would labor without respite until, famished, I refused to core any more. One evening toward the end of the week I finally asked the guys about the unceasing pace, and at last the truth came out. All three of my colleagues admitted to being re-lieved when I demanded that we eat each day, because none of them wanted to be the first to let on that he was starving. Apparently, before I, the first woman to join them on a campaign, came along and insisted on sense, their testosterone-driven stubbornness resulted in whole days of intense physical effort in the field, entirely without food. Thereafter, the confessions got even better. It turned out that none of them wanted to be the first to concede that he was cold at night, and so they slept in their shared room with the windows wide open even when temperatures in the mountains dropped to near freezing. At that mo-ment, I was very happy with my gender. As a woman scientist, I got 99 prob-lems, but at least starving or freezing to death to protect my ego ain't one.

▨ The research focus of the WSL dendroclimate team was to use tree rings to reconstruct climate over past centuries. To study past climate prior to the early twentieth-century start of the *instrumental climate record*—the meteorolog-ical data derived from daily measurements at weather stations around the world—we make use of *paleoclimate proxies*. These biological or geological archives, such as the layers in ice cores, lake sediments, trees, and corals, record climate conditions and can therefore be used as sources of climate information. Tree-ring records provide good value for money in that context. They are relatively easy and cheap to acquire and analyze. With trees and forests covering large swaths of the earth's surface, tree-ring data are some of the most commonly used climate proxies, especially for the most recent 1,000 to 2,000 years, the period with the most dendrochronological data.

In that context, our Lake Gerber field campaign was aimed at developing a millennium-long climate reconstruction for the Pyrenees. With a 1,000-plus-year tree-ring chronology in hand, our odds for achieving this goal initially looked good. But as we discovered, mountain pine trees in the Pyrenees are sensitive to a mix of limiting climate factors. The trees grow at high elevation and are thus limited by cold temperatures that persist even in summer. But they also grow in the seasonally dry Mediterranean region and so are also limited by a lack of summer rain. As a result, the pines form narrow rings both when it is cold and when it is dry. Their narrow rings can indicate cold summers or dry summers, and their yearly ring width cannot be used to reliably reconstruct either temperature or precipitation independently. It was looking as if our tree-ring data were not going to pay off in a Pyrenees climate reconstruction despite our hard work. Luckily, we can also measure other parameters in tree rings, apart from their width. Using tree-ring radiometry, for instance, we can measure the wood density of individual rings, which often captures summer temperature variations better than the ring's width. The maximum density of the latewood portion of a ring, in particular, reflects how much the cell walls in that ring have thickened by the end of the growing season. This in turn is strongly determined by the temperature that the tree has experienced during its growing season: trees form denser latewood in hot summers than in cold summers. Therefore, *maximum latewood density* provides a very good record of summer temperature in the year that the ring was formed, and wood-density measurements can be used as proxies for past summer temperatures. Like ring-width measurements, tree-ring-density measurements are absolutely dated and give a data point each and every year. When we measured maximum latewood density in our Pyrenees tree-ring chronology, we found that it was strongly and uniquely influenced by summer temperature, which allowed us to use the chronology to reconstruct past summer temperatures after all.

The concept behind using tree rings to reconstruct past climate is fairly straightforward. The width (or density) of the annual rings can be measured, absolutely dated, and compared on a year-by-year basis with instrumental weather-station data. In our Pyrenees project, we collected cores in the summer of 2006, so the last fully formed ring on the cores was the 2005 ring. The oldest sample in the Pyrenees tree-ring chronology dates back to 924 CE, and at least five samples date back to 1260 CE. Because the replication of five samples is more robust, we consider the year 1260 to be the beginning of our reconstruction. Fortunately, instrumental temperature measurements started early in the Pyrenees, and temperature data from the nearby mountain observatory Pic du Midi are available from 1882 onwards. We can thus compare our maximum-latewood-density data for each year with the summer-temperature data for the same year collected at the observatory for the period 1882–2005. As a result, we have 124 years of tree-ring data to compare with 124 years of summer-temperature data. Given that our maximum-latewood-density measurements are good recorders of summer temperature, we can keep it simple and link maximum latewood density (MXD) for each year to summer temperature (Tsummer) for that year in a linear equation or model:

$$\text{Tsummer}(t) = a^*\text{MXD}(t) + b.$$

This equation states that summer temperature in year t can be expressed as a function of maximum latewood density of the ring formed in that year. Because density measurements are made in grams per cubic centimeter (g/cm^3) but we want to reconstruct summer temperature in degrees Celsius, we need the constants a and b to transform g/cm^3 into degrees Celsius. We use the MXD and Tsummer data for the period 1882–2005 to calculate the values of a and b. To test the strength of the relationship between MXD and Tsummer, we calculate how many of the 124 years of overlap feature hot summers (high Tsummer) corresponding to high density (high MXD), and vice versa. If the relationship is strong, then we can use this same equation, with the same a and b values, to calculate summer temperature for any year for which we have a maximum-latewood-density value, all the way back to 1260. We just multiply maximum latewood density for a certain year by a and add b to it, and this will give us an estimate of summer temperature for that year. By doing this for each year of the tree-ring record, we can reconstruct summer temperature prior to the start of the instrumental record, back to 1260.

The simplest models (or equations), such as the one above, use one tree-ring chronology to predict and reconstruct one instrumental climate time series from a nearby location. Often such simple models can be improved by combining tree-ring chronologies from multiple locations. In the Pyrenees, for instance, our Lake Gerber maximum-latewood-density record better represented the instrumental summer temperature if it was combined with a density record from a nearby timberline site in Sobrestivo, about 45 miles west. The model can also be optimized by selecting the climate variable that most strongly influences tree growth. Maximum latewood density of mountain pine in the Pyrenees, for instance, is more sensitive to temperature variability in May, August, and September than to that in June and July. In other words, our tree-ring data gave us a more reliable estimate of past May–September temperatures than of June–July temperatures. A variety of climate variables (e.g., rainfall versus temperature, one month versus another, instrumental data from a single meteorological station versus the average of data from many stations in a region, and so forth) can thus be used on the lefthand side of the equation, and a variety of tree-ring data (e.g., tree-ring width and/or density, data from one tree species or multiple species, data from one site or multiple sites) on the righthand side. An important part of a dendroclimatologist's job is to select the right tree-ring data and climate data for reconstruction and to determine through statistical analyses which combination gives the most reliable and robust results.

In 1998, climatologist Michael Mann, paleoclimatologist Ray Bradley, and dendrochronologist par excellence Malcolm Hughes took this simple concept and moved it a giant step forward. By the late 1990s, the extraordinary character of twentieth-century global warming was unmistakable, and Mann, Bradley, and Hughes aimed to put the recent warming in a historical context in order to find out whether it could be part of a natural climate cycle. For this purpose, they developed a year-by-year reconstruction of Northern Hemisphere temperatures for the past 600 years. They combined tree-ring data with ice-core data and other proxies to reconstruct yearly temperature variations averaged over the Northern Hemisphere. They applied a new statistical approach to pro-

The Hockey Stick

Northern Hemisphere temperature relative to average temperature (1961–1990)

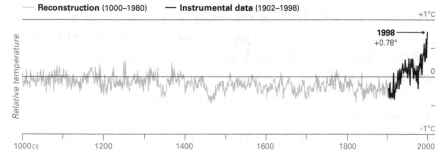

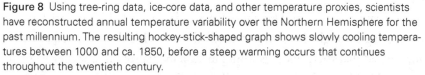

Figure 8 Using tree-ring data, ice-core data, and other temperature proxies, scientists have reconstructed annual temperature variability over the Northern Hemisphere for the past millennium. The resulting hockey-stick-shaped graph shows slowly cooling temperatures between 1000 and ca. 1850, before a steep warming occurs that continues throughout the twentieth century.

duce a single reconstruction for the entire hemisphere that covered the period back to 1400 CE, which was published in the scientific journal *Nature*. Their reconstruction demonstrated that twentieth-century global warming was unprecedented over the past 600 years. In a follow-up paper a year later, they extended the reconstruction even further back in time to 1000 CE. The key figure of their paper was a plot of Northern Hemisphere temperature change over time that resembles a hockey stick (fig. 8). The graph shows slowly cooling temperatures between 1000 CE and about 1850 CE (the hockey stick's "shaft") before a steep warming that continued throughout the twentieth century (the "blade"). The hottest year of the 1,000-year-long Hockey Stick was 1998, also the most recent year of their record.

The Mann, Bradley, and Hughes Hockey Stick paper was the first to show that twentieth-century warming was unprecedented in a 1,000-year perspective and thus unlikely to be part of a natural cycle. Given the relevance of this finding, the Hockey Stick graph was featured prominently in the 2001 report of the Intergovernmental Panel on Climate Change (IPCC). The IPCC is a United Nations scientific body entrusted with providing a comprehensive, scientific, and objective overview of climate change and its impacts on society. It won the 2007 Nobel Peace Prize, jointly with former US vice president Al Gore. The IPCC does not carry out original research, but every five years or so it produces a hefty report based on the published scientific literature. The report is written by volunteering scientists and reviewed by governments before publication. The 2001 IPCC report was an 800-page tome. Because no sane policymaker carries 5.5 pounds of paper around in her briefcase (or has time to read such a lengthy report), a handy, approximately 30-page summary for policymakers was published that highlighted the most important findings with a handful of graphs. The Hockey Stick graph was prominent in this summary, and it subsequently drew global media attention when a large poster of it was used as the backdrop to a television announcement presenting the 2001 IPCC report.

As Mann, Bradley, and Hughes's 1998 and then 1999 Hockey Stick papers were going through the review process, the three highly respected scientists anticipated that their findings would make a media splash, but they were unprepared for the media frenzy that followed. The Hockey Stick story was picked up by all the major media outlets, and in a *New York Times* interview Mann reemphasized that "the warming of the past few decades appears to be closely tied to emission of greenhouse gases by humans and not any of the natural factors."*

What followed was almost two decades of relentless political inquisition and intimidation. Two of the first and foremost political mudslingers were James Inhofe, a US senator from Oklahoma and chair of the Senate Committee on Environment and Public Works, and Joe Barton, a congressman from Texas and chair of the House Energy Committee. Inhofe is well known for repeatedly calling man-made global warming "the greatest hoax ever perpetrated on the American people." In their sustained efforts to oppose potential restrictions on greenhouse-gas emissions and to dismiss the ideas behind anthropogenic climate change, both Republican politicians honed in on the Hockey Stick, the poster child of the IPCC and of climate-change policy.

In the years 2003 through 2006, Inhofe and Barton convened multiple congressional hearings to which they invited the Hockey Stick scientists and, more pointedly, a wide range of climate-change skeptics. Debate between scientists,

---

*William K. Stevens, "New evidence finds this is warmest century in 600 years," *New York Times*, 28 April 1998.

in which controversial methods and results are double-checked and the validity of conclusions argued, is a crucial part of the scientific process, but the political arena is not the place for such debate. Scientific facts are not decided by a majority vote. Or as Sherwood Boehlert, the chairman of the House Science Committee, himself a conservative Republican, phrased it in a letter to Barton: "My primary concern about your investigation is that its purpose seems to be to intimidate scientists rather than to learn from them, and to substitute congressional political review for scientific review."* This politicization of the Hockey Stick hit rock bottom in September 2005, when Senator Inhofe invited Michael Crichton, the creator of the popular TV series *ER* and author of fictional thrillers such as *Jurassic Park*, to testify before the Senate on the legitimacy of climate change during a hearing on the role of science in environmental policy making. Inhofe referred to the novelist Crichton as a scientist and made his fictional thriller *State of Fear* required reading for the Senate Committee on Environment and Public Works. In his novel, Crichton imagines a world where climate change is not an ecological reality but rather an evil eco-terrorist conspiracy. In his two-hour-long testimony before the committee, Crichton expressed his strong doubts about "whether the methodology of climate science is sufficiently rigorous to yield a reliable result."† He then sat back as Senator Hillary Rodham Clinton expressed her opinion that his views "muddied the issues around sound science." In my view, a fiction writer as a star witness before a US Senate committee on the validity of scientific research seems just as unbelievable as a T-Rex running amok in a Disney-esque theme park.

In an additional attempt to break the Hockey Stick, Inhofe's greatest ally, Barton, demanded from the three scientists full records of all of their climate-related research. The laundry list of demands included exhaustive information about all financial support they had received during their long careers, the source of funding for every study they had conducted, and all data and code for every paper they had ever published. In a letter asking Barton to withdraw his requests, Democratic congressman Henry Waxman wrote, "These letters

---

*Boehlert to Barton, 14 July 2005, https://www.geo.umass.edu/climate/Boehlert.pdf.
† "The role of science in environmental policy making," Hearing before the Committee on Environment and Public Works, US Senate, 28 September 2005, https://www.govinfo.gov/content/pkg/CHRG-109shrg38918/html/CHRG-109shrg38918.htm.

do not appear to be a serious attempt to understand the science of global warming. Some might interpret them as a transparent effort to bully and harass climate change experts who have reached conclusions with which you disagree."* The goal of Inhofe, Barton, and their disciples in this "misguided and illegitimate investigation,"† according to the Republican Boehlert, was to ensure that legislation to regulate greenhouse-gas emissions in the US never passes. As reported by Naomi Oreskes and Erik Conway in their 2010 book *Merchants of Doubt*, this tactic, which keeps a controversy alive by spreading doubt and confusion despite a scientific consensus, has successfully been applied in the past by the tobacco industry to deny the connection between smoking and cancer.

The ruthlessness of the climate-change deniers' tactics became very evident in November 2009, when hackers broke into the server of the Climate Research Unit at the University of East Anglia and stole and made public thousands of researchers' private email correspondences. The emails showed some researchers to be rude, others cocky or petty. What they did not show was a sweeping scientific conspiracy perpetrated by a shadowy cabal of global proportions. Yet, that is what the climate-change-denial league behind the hacking claimed. It is no coincidence that this illegal hacking occurred just weeks before the UN Climate Summit in Copenhagen, at which an international framework for climate-change mitigation was to be established. The media were quick to dub the hacking incident "Climategate," focusing not on the hacking crime itself but on the language used in the emails. The Climate Research Unit—backed by multiple professional scientific organizations—rejected the accusations. No fewer than eight independent committees, including one from the US Environmental Protection Agency, have investigated the emails and allegations, and all came to the same conclusion: that there was no evidence of fraud or scientific misconduct. Nevertheless, Climategate was very effective in drawing media attention away from the important goals of the Copenhagen climate summit, while dragging some of the world's most renowned climate scientists through the mud. Mann, Bradley, and Hughes are still deal-

---

* Henry A. Waxman to Chairman Barton, 1 July 2005, https://www.geo.umass.edu/climate/Waxman
  .pdf.
† Boehlert to Barton, 14 July 2005.

ing with the aftermath of the Hockey Stick controversy and Climategate, almost two decades after their paper was first published.

Over the past twenty years, more and more scientific studies showing the unprecedented nature of the current climate have confirmed and advanced the concept of the original Hockey Stick paper. As time goes on, reality is also catching up with us. Nineteen ninety-eight, the most recent and hottest year of the Hockey Stick record, is now only the tenth hottest year on record, with nine even hotter years occurring since. Needless to say, the relentless harrassment inspired by those who stand to lose much if actions are taken to limit emissions has required an inordinate amount of my colleagues' attention, time, and energy. Attention, time, and energy that they were not able to spend on research, on sampling more and older trees, on crossdating more samples, or on publishing more scientific results. And therein seems to lie a key motivation for this dogged political inquisition and intimidation: to keep climate scientists from doing their job—studying natural and man-made climate change and sharing their findings with the world.

Seven
Wind of Change

The first more or less reliable thermometer was invented in 1641 by Ferdinand II de' Medici, the grand duke of Tuscany and a student of Galileo Galilei's. Flushed with his success, Ferdinand and his brother set up a network of 11 meteorological stations in Italy and neighboring countries. The stations were operated from 1654 onwards by monks and Jesuit priests who for years read the thermometers every three to four hours. But in 1667, most of this early network was shut down by the Catholic Church on the premise that only the Bible, not instrumental readings, could be used to interpret nature; only 2 stations continued until 1670. Happily, temperature measurements in central England started in 1659, only five years after de' Medici's efforts began in earnest, and they have persisted through the predations of time ever since. The resulting instrumental record for central England is the longest continuous sequence of temperature measurements in the world. In the US, measurements did not start until 1743, in Boston. In the Southern Hemisphere, only one record predates 1850: that for Rio de Janeiro, where measurements started in 1832. It was not until the early twentieth century that a worldwide network of reliable temperature measurements became available, and even for the twentieth century there are big geographical gaps in the network. For instance, the temperature and precipitation records from Kigoma that Kristof and I transcribed by hand during our Tanzania field campaign only started in 1927. The instrumental climate record is even further complicated by the fact that we only started measuring the climate on a global scale right about the same time that we started interfering with it. By the time a global meteorological network was set up in the early twentieth century, the Industrial Revolution—and with it the ever-increasing burning of fossil fuels and emission of greenhouse gases into the atmosphere—was already well under way.

Greenhouse gases, such as carbon dioxide, trap heat and prevent it from escaping into space. It is as if a carbon-dioxide miasma were surrounding the earth, heating it up and growing ever denser with the burning of more fossil fuels. Starting with the Industrial Revolution in the late eighteenth century, this has led to an enhancement of the natural greenhouse effect and rising temperatures at the earth's surface, a.k.a. global warming. The proxy record provided by ice cores from the Antarctic has allowed us to put this increase in carbon-dioxide concentrations in the atmosphere into a context of almost a million years. Data gathered from coring deep into the Antarctic ice sheet and measuring the amount of carbon dioxide in the air bubbles inside the increasingly old layers of ice, tell us that today's concentration of carbon dioxide in the atmosphere is almost 40 percent higher than at any time in the previous 800,000 years. Because most meteorological stations were established only after the start of the Industrial Revolution, the climate we have been recording with our instruments is under an anthropogenically enhanced greenhouse effect. We don't have an instrumental record of what the climate was like in its more natural state, before we started turning the atmosphere into a greenhouse. We need paleoclimate proxies to understand what the "natural" climate, one without large-scale interference by humans, was like.

■ Paleoclimate proxies have taught us that the earth's climate is a complex system that demonstrates inherent variability and that in addition to artificial changes in atmospheric greenhouse-gas concentrations, it also reacts to changes in the earth's orbit, the sun's radiation, and volcanic activity. When the earth changes its position relative to the sun due to shifts in its elliptical orbit around the sun or in the tilt of its axis, the amount of solar radiation that reaches the earth changes. With the sun being the primary source of the earth's heat, such orbital changes result in changes in the global temperature. Orbital variations are cyclical in nature, and they are slow: they affect the earth's climate over periods of 100,000, 40,000, and 20,000 years. Though prolonged, they are powerful; they have such a strong impact on the earth's temperature that they are responsible for the occurrence of ice ages. Cold ice ages (or *glacials*) alternate with warmer periods, or *interglacials*, on timescales of ca. 100,000 years, and this regular and repeated alternation is beautifully recorded in ocean sediment and Antarctic ice-core records. We are currently in an interglacial period, the

Holocene, which started around 11,650 years ago. With interglacials lasting anywhere between 10,000 and 50,000 years, we are inevitably orbitally bound to return to an ice-age climate in the future. However, the recent enhanced greenhouse-gas effect and the global warming it causes may well overhaul our million-year-long ice-age history.

In addition to orbital changes, the amount of radiation that the sun itself emits can also change over time and impact Earth's temperature. The amount of radiation given off by the sun varies in cycles, which last from decades to centuries—much shorter than the cycles in Earth's orbit. Solar radiation creates *isotopes* in the earth's atmosphere, alternate forms of chemical elements that differ in relative atomic mass but not in chemical properties. For instance, beryllium-10 (Be10), the radioactive isotope of beryllium (Be9) with a half-life of more than a million years, is created by powerful outbursts of solar radiation. Atmospheric Be10 is deposited in the air bubbles in the snow and ice layers of Greenland and Antarctica, and Be10 peaks in dated ice cores can be used as a proxy for the sun's past activity and cyclicity. Fluctuations in the sun's radiation can also be estimated based on sunspots—regions of reduced temperature—on the sun's surface. Fewer visible sunspots mean that the sun is less magnetically active and that less radiation is sent to the earth. Sunspots sometimes are large enough to be seen by the naked eye, and early modern scientists used telescopes to observe sunspots starting in the 1610s. The 400-plus years of sunspot observations have provided a documentary solar-radiation proxy revealing a regular 11-year cycle in the number of sunspots and related solar radiation. It was these cycles that Douglass was trying to trace when he first started looking at tree rings. The cycles, however, have only a subtle impact on the earth's climate. Of larger import are the longer, multi-decade periods when sunspot activity is depressed, for instance, during the solar Maunder Minimum, named after the solar astronomer couple Annie and Edward Maunder, who were contemporaries of Douglass's. For the 70 years from 1645 to 1715, astronomers observed far fewer sunspots on the sun's surface than at any point before or after. Incidentally, this 70-year Maunder Minimum corresponds almost perfectly with the reign of Louis XIV, the Sun King, over France (1643–1715).

Volcanic activity is the third major driver of natural climate variability. When a large volcano erupts, and especially when it does so explosively, it can eject large amounts of aerosols, dispersing fine particles such as sulfur dioxide

(SO_2), as well as ash, into the atmosphere. Over the course of a few weeks to months, this sulfur dioxide is converted into sulfuric acid (H_2SO_4), and once formed, the sulfuric aerosols are widely dispersed throughout the stratosphere (the upper region of the atmosphere), where they can stay for years. Such volcanic aerosol veils can stop some of the sun's radiation from reaching the earth's surface, resulting in cooler temperatures. Volcanic aerosols' effects are therefore the opposite of greenhouse gases'; rather than warming the earth's surface, volcanic dust particles block the sun's radiation, cooling the earth's temperatures for up to two years following an eruption. The eruptions of tropical volcanoes, whose aerosols are spread most easily throughout the entire stratosphere, typically have a more global-scale impact on climate than do the eruptions of volcanoes at higher latitudes, and strong eruptions will have a broader impact than weaker eruptions. The cooling effect of volcanic eruptions on the earth's climate may be short-lived—a few years at most—but it can be drastic. When Mount Pinatubo, close to the equator in the Philippines, erupted in June 1991, it ejected an ash cloud that rose 22 miles into the air, deep into the stratosphere. During the 15 months following the Mount Pinatubo eruption, average global temperatures dropped by about 1 degree Fahrenheit. The abrupt cooling caused by the eruption was picked up by temperature-sensitive tree-ring records across the globe, demonstrating that these records can be used to determine the year and the magnitude of past volcanic eruptions and to analyze their climatic impact.

Our understanding of how these three drivers—changes in the Earth's orbit, in solar radiation, and in volcanic activity—have combined to impact our past climate is most advanced for the past approximately 1,000 years. The concept of natural climate variability over this period was introduced in a plot created by the English climatologist Hubert Horace Lamb in 1965 (fig. 9). Lamb's plot depicts 1,000 years of temperature variations in Central England, including a climate transition that occurred as medieval Europe advanced into the Renaissance and the Age of Discovery. Temperatures were relatively warm during what modern scientists have termed the *Medieval Climate Anomaly* (Lamb called it the *Medieval Warm Period*), ca. 900–1250, but cooled down substantially during the subsequent *Little Ice Age*, ca. 1500–1850, a period when more volcanoes erupted, the sun lost some of its power (for instance,

The Noodle
900–1965CE

— **Hubert Horace Lamb's estimate, 1965**

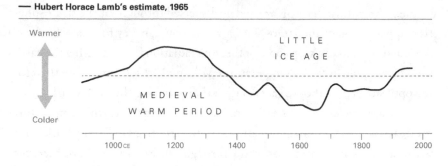

The Hockey Stick
Northern Hemisphere temperature relative to average temperature (1961–1990)

— **Reconstruction** (1000–1980)   — **Instrumental data** (1902–1998)

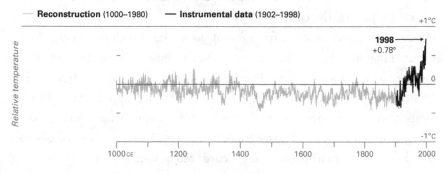

The Spaghetti Plate
Northern Hemisphere temperature relative to average temperature (1961–1990)

≡ **Various reconstructions** (700–1995)   — **Instrumental data** (1856–2005)

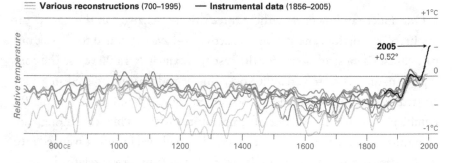

Figure 9 Hubert Horace Lamb's sketch showing climate variability over the past thousand years includes a plot (*adapted above, top*) that shows warm temperatures during medieval times, a cool Little Ice Age, and warming temperatures again into the twentieth century. Scientific visualizations of a thousand years of global temperature change have matured. David Frank called these developments an evolution from "a noodle, to a hockey stick, to a spaghetti plate."

during the Maunder Minimum), and the proportion of solar energy entering and leaving the Earth system shifted. Unlike "real" ice ages, the Little Ice Age was not caused by orbital changes, and it was much less severe, less incessantly cold, and less uniform across the Earth's surface. Lamb's Little Ice Age ended in the mid-nineteenth century, as the advance of industrialization led to a steady rise in temperatures.

Half a century of paleoclimate research has brought nuance to Lamb's pioneering image of a thousand years of temperature change. My Pyrenees companion David Frank called our progress the evolution from "a noodle, to a hockey stick, to a spaghetti plate." Lamb's graph does look like a cartoon noodle, yet it remained the go-to image of temperature history until the late 1990s, when the more data-rich and much more computationally heavy Hockey Stick took its place. After the political controversy caused by the Hockey Stick, many more research groups started reconstructing global temperatures back in time, throwing ever more data, ever more computational power, and an amalgam of methodological advances at the challenge. The result is a "spaghetti plate" of temperature reconstructions that portray some periods similarly—such as a warm eleventh century and unprecedentedly warm twentieth century—but show a wider range of possibilities, and thus quite a bit of uncertainty, for other centuries.

While we were colleagues at the WSL, David involved me in a project that attempted to bring order to this spaghetti plate of reconstructions by subjecting each of the individual spaghetti strands (or temperature reconstructions) to all possible combinations of methods, resulting in an ensemble of more than 200,000 reconstructions in total. At first sight, this approach would appear to make the tangle of lines even more complex, but there was a method to our madness. A pattern emerged from the Spaghetti Plate that allowed us to estimate the most likely temperature variations over the past millennium. We found that the most recent period of the reconstruction ensemble, which ended in the year 2000, was most likely around 0.5 degrees Fahrenheit warmer than the warmest period of the Medieval Climate Anomaly* and around 1.3 degrees Fahrenheit warmer than the coldest period of the Little Ice Age.[†] The recon-

---

* 1071–1100 CE.
† 1601–30 CE.

struction ensemble shows us that medieval times were warmer than the Little Ice Age but not nearly as warm as it is today. It is worth noting that every single year since 2000 has been warmer than the year 2000, the last year of the reconstruction. As a result, the current most recent period (ending in 2016) is another 0.9 degrees Fahrenheit warmer than the warmest period of the ensemble reconstruction.* The temperature differences in the study—0.5 degrees, 0.9 degrees, 1.3 degrees—might seem small, but I was overwhelmed when I first looked them up. I was stunned to realize that the Earth has warmed more in the past 17 years (2000–2016) than it had cooled over the 500 years from the Medieval Climate Anomaly into the Little Ice Age.

Moreover, the warming in recent decades is undeniably global in scale—no matter which Spaghetti Plate reconstruction you look at, it is always there—whereas the transition from the Medieval Climate Anomaly into the Little Ice Age was not uniform in place or in time. The Little Ice Age started much earlier in the Arctic region (ca. 1250 CE), for instance, than in lower latitudes, such as the European Alps (ca. 1500 CE), and while the Little Ice Age brought colder temperatures to most areas, in some areas it was defined by wetness rather than coldness. In the Atlas Mountains of Morocco, on the very northwestern tip of the African continent, for instance, the Little Ice Age started around 1450 CE. It is recorded as a wet period in the tree rings of a grove of 500-plus-year-old Atlas cedars, whose yearly growth is limited by how much moisture is available to them.

I have not visited the grove of Atlas cedars myself, but it is on my bucket list. The Moroccan Atlas cedars are some of the oldest trees in Africa, with clear annual rings that crossdate well and are reliable recorders of drought. From a dendrochronologist's point of view, those are very desirable characteristics, and many research teams have visited and cored the cedars over the years. So far, however, no one has brought the 40-inch borer needed to reach the pith of these mastodons, which can be up to 10 feet in diameter. When Jan Esper and his team sampled the trees in 2002, long before I started working with them in the Pyrenees and in Greece, they extracted a core from one tree

* This calculation is based on the difference in GIStemp (Goddard Institute for Space Studies Surface Temperature Analysis) Northern Hemisphere mean annual temperatures between 1987–2016 and 1971–2000.

that counted 1,025 rings, but their 24-inch borer did not reach the pith of the tree and so its oldest rings remain unextracted and unexamined. According to Jan, some of the trees could easily be 1,300 to 1,400 years old and could be used to extend the millennium-length Atlas cedar tree-ring chronology back to before medieval times, if we could just reach their inner rings.

The Atlas cedars are limited by spring drought: they are happy with fat rings when it's wet, unhappy with narrow rings when it's dry. Their tree-ring chronology functions as a 1,000-plus-year-long reconstruction of Moroccan drought. The earliest (ca. 400) rings of the trees are remarkably narrow as they recorded a severe and long-lasting medieval drought. From ca. 1450 onward the trees received much more moisture, until ca. 1980, when another severe drought set in. This recent drought, which is still ongoing, has impacted regional agriculture and tourism and is threatening the cedar forests. These forests have suffered from overexploitation, overgrazing, and repeated burning for centuries and were already in bad shape before the recent 30-plus-year drought. The drought has been the final blow for many Atlas cedars, which are now listed as endangered on the IUCN* Red List of Threatened Species.

When I first started working with Jan at the WSL, around the time of our Pyrenees field trip, his idea was to use the Moroccan drought reconstruction as a base to develop a "Drought Hockey Stick." Jan wanted to test whether Northern Hemisphere drought variability over the past thousand years could be captured in a single graph, similar to the iconic Hockey Stick but illustrating hemispheric-scale trends in precipitation, rather than temperature. Such a Drought Hockey Stick did not exist at the time and does not exist now, because rainfall and drought vary much more over space than temperature does, and are harder to capture by calculating averages.

If you compare, for instance, the year-to-year variability in annual temperature in Meknes, a meteorological station in the Atlas Mountains, with that in Algiers, 600 miles to the northeast on the Mediterranean coast, you'll find that they are very similar.[†] Years that are hot in Meknes are typically also

---

* International Union of Conservation of Nature and Natural Resources.

[†] The Pearson correlation coefficient for the two temperature time series is positive and strongly significant ($r = 0.66$, $p < .001$, 1961–2016).

hot in Algiers. Cold years in Meknes are typically also cold in Algiers. On the other hand, the year-to-year variability of annual rainfall in Meknes is virtually unrelated to that in Algiers.* Years that are wet in Meknes can be dry, average, or wet in Algiers—there is no relationship. To draw a picture of large-scale temperature trends over time, as with the Hockey Stick, it can make sense to average temperature data from sites that are 500 miles or more apart, because they show the same variability. Averaging precipitation or drought data over such large distances, however, makes less sense: averaging distant data that are unrelated to one another will create a flat line that doesn't provide much information at all. Taking this into consideration, Jan proposed to start with a smaller geographical range—a European Drought Hockey Stick—rather than one of hemispheric scale, but we knew that even that would be a challenging aim. Adding to this challenge for me were certain aggravating circumstances: (*a*) I had never studied European climate before and (*b*) I had never studied paleoclimate before.

Before arriving at the WSL, I had studied tree rings in sub-Saharan Africa and in the Californian Sierra Nevada. Neither of these projects had had anything to do with European climate or even with climate reconstruction. Needless to say, I was in over my head, but I did not want to show it. I was now part of a research group that could not even admit to being hungry in the field, let alone to being ignorant. In our daily and weekly lab discussions, Jan, David, Ulf, and our colleague Kerstin Treydte would throw the terms *Medieval Climate Anomaly* and *Little Ice Age* around in their casual banter as if they were Bundesliga teams. I remember covertly looking up the terms on Wikipedia in an attempt to keep up. What I really wanted to do was hang a big sign over my desk that read:

MEDIEVAL CLIMATE ANOMALY = 900–1250 CE = WARM
LITTLE ICE AGE = 1500–1850 CE = COLD

but that would have been a dead giveaway. Having to Wikipedia key concepts of one's supposed field of study does not help with overcoming impostor syn-

---

*The Pearson correlation coefficient for the two precipitation time series is low and insignificant ($r = 0.17, p > .1$, 1961–2016).

drome. To add insult to injury, I also had to ask my colleagues some very basic questions about European climate, such as what the main driver of European climate variability is. In the places where I had worked before, year-to-year climate variability is mainly influenced by the El Niño Southern Oscillation (ENSO) system, a.k.a. El Niño. I knew that this Pacific ocean-atmosphere interaction pattern did not have a strong influence on European climate, but I had no idea what governed it instead. In retrospect, I admire my colleagues for not shouting: "It's the NAO, stupid!"

The NAO, or *North Atlantic Oscillation*, is a seesaw (or oscillation) in the air pressure between two major pressure centers over the North Atlantic Ocean, the Azores High and the Icelandic Low (fig. 10A). Air pressure (or atmospheric pressure) is important because it is linked to weather patterns: low pressure fields typically lead to cloudy, windy, and rainy weather, whereas high pressure leads to calm and sunny weather. Intuitively, it makes sense that the atmospheric pressure over the sunny Azores Islands, off of Portugal, is almost always higher than it is over rainy Iceland. But the difference in pressure between the Azores High and the Icelandic Low is larger in some years than it is in others. The pressure difference is very large during *positive* phases of the NAO, when the Icelandic Low is even lower than normal and the Azores High is even higher than normal. Positive NAO years represent one position of the seesaw, but the position switches during *negative* NAO phases, when the pressure difference between the two is small.

The Azores High is an *anticyclone*, which swirls wind in a clockwise fashion from the tropics toward the North Atlantic Ocean and then toward Europe. The Icelandic Low *cyclone* moves air around in the opposite, counterclockwise direction, bringing air from the Arctic toward the North Atlantic Ocean and then to Europe. The two pressure centers function as cogs in the North Atlantic climate wheel; both turn at full speed during positive NAO phases, when warm air is propelled from the North Atlantic Ocean toward Europe. The strong Icelandic Low brings wet and stormy weather to the British Isles and Scandinavia; the strong Azores High brings drought to the western Mediterranean, and the strong winds bring warm and mild weather to central Europe. The opposite happens during a negative NAO phase, when

The North Atlantic Wind Machine

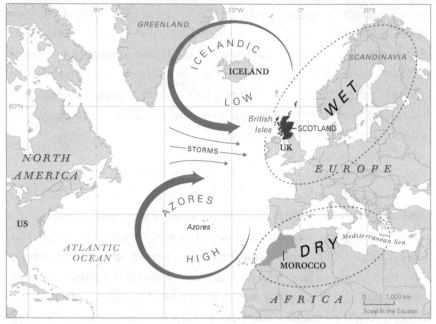

Figure 10A Europe's weather largely depends on two atmospheric-pressure centers over the North Atlantic: the Azores High and the Icelandic Low. Together, they function as cogs in a giant wind machine. When the pressure difference between them is large, winds swirl at full speed, bringing storms to the British Isles and Scandinavia, drought to the western Mediterranean, and mild weather to central Europe. When the pressure difference is small, they rotate slowly, keeping warm North Atlantic winds from reaching Europe. The British Isles are then drier, and the western Mediterranean is wetter than normal.

both the Azores High and the Icelandic Low are weaker than normal. In negative NAO years, the British Isles are drier than normal, which is still pretty wet, and the western Mediterranean is wetter than normal, which is still pretty dry. The cogs in the North Atlantic wind machine turn slowly, keeping warm North Atlantic winds from reaching Europe and clearing the way for cold air from the northeast to move in.

Only vaguely aware of any of these major aspects of the European climate system, I nonetheless embarked on my journey to develop a Drought Hockey Stick for Europe. With Jan's thousand-year Moroccan drought record in hand, I looked for European drought reconstructions of roughly the same length to

Evidence in Trees and Stalagmites

1049–1995CE

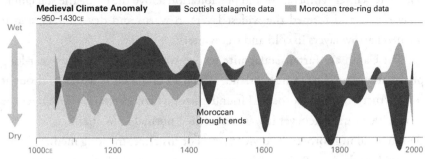

Medieval Climate Anomaly ~950–1430CE   ■ Scottish stalagmite data   ■ Moroccan tree-ring data

Wet

Dry

Moroccan drought ends

1000CE   1200   1400   1600   1800   2000

Figure 10B Stalagmites can provide a proxy of winter climate. Comparing a thousand years of stalagmite data from Scotland with tree-ring data from Morocco reveals an inverse relationship: whenever it was wetter than normal in Scotland, it was drier than normal in Morocco, and vice versa.

compare it with.* The first hurdle I encountered was the shortage of such reconstructions. Because of its long history of intensive human wood use, the European continent generally lacks old trees. Its oldest dendrochronologically dated tree, Adonis, is barely over 1,000 years old. And like Adonis, most of Europe's old trees grow in remote places that are hard for humans to reach, such as high up in the mountains, so that their tree rings typically record changes in temperature, not drought. I therefore had to extend my search to include proxies other than tree rings, pushing myself even further outside my comfort zone. Of the handful of European drought reconstructions for the past millennium, only one matched the Moroccan Atlas cedar record: a stalagmite record from a cave in Scotland. Just like tree rings, stalagmites in caves can form growth layers. In the Uamh an Tartair cave ("Roaring Cave"), in northwestern Scotland, these growth layers are annual: one stalagmite layer represents the stalagmite's growth over one year. Andy Baker, a geoscientist at the University of New South Wales, and his team collected a small stalagmite from the cave. At the time of its collection, the stalagmite was only 1.5 inches high, but it was still actively growing and adding layers. The researchers counted 1,087 growth bands in the polished

---

* Technically speaking, Morocco obviously is not part of the European continent, but its location on the northwestern tip of Africa represents climate variability for the southwestern Mediterranean region.

stalagmite and found that the width of these bands, like those in tree rings, was related to winter temperature and precipitation above the cave. The stalagmite grew fastest and formed the widest layers in warm and dry winters, and it formed narrow layers in cold and wet winters.

The Uamh an Tartair stalagmite record was thus a 1,000-plus-year-long proxy of winter climate in Scotland. When I compared it with the drought reconstruction from Morocco, I found a strong inverse relationship: over the past 1,000 years, whenever it was wetter than normal in Scotland, it was drier than normal in Morocco, and vice versa. For instance, the long medieval Moroccan drought (ca. 1025–1450) corresponded to a period of even wetter than normal conditions in Scotland, and as Morocco came out of its drought around 1450, Scotland got drier (fig. 10B).

When I first told Jan about the Scotland-Morocco seesaw I had found, he was dismissive. Dendrochronologists habitually distrust other climate proxies, such as stalagmites—working with tree rings spoils us. We can take many samples from a site, which allows us to crossdate and double-check our dating. We have a good, mechanistic understanding of how tree growth is related to climate. We have one tree ring and one data point for every single year, so that we can check through direct comparison whether our tree rings are a good proxy for instrumental climate variability. Not many other climate proxies have all these advantages. "One measly stalagmite, that's all you got? And its dating might be a couple of years off but you're not sure because you don't have anything to crossdate it against? Pfft! I scoff at your stalagmite!" Those weren't Jan's exact words, but you get the picture.

But when I unveiled the plot of 1,000 years of Scottish pluvials coinciding with Moroccan droughts, and vice versa, Jan changed his mind. That was the moment when he and I both realized that we were looking at something important, something science had not yet discovered. We were seeing 1,000 years of not only the seesaw between Scotland and Morocco but also the seesaw of the NAO. The Uamh an Tartair stalagmite is a proxy not only for precipitation in Scotland but also for the Icelandic Low. Likewise, the Atlas cedar tree-ring record is a proxy not only for drought in Morocco but also for the Azores High. By combining proxies for those two cogs in the NAO climate machine, we had created a 1,000-year-long reconstruction of the NAO. We had set out to develop a regional Drought

Hockey Stick and instead uncovered the history of one of the most influential global-climate phenomena. The tiny, 1,000-year-old rings that we had measured in the Moroccan cedars were telling us stories not only about past droughts and pluvials but also about the much larger atmospheric mechanisms that propelled them and the complex global climate machine as a whole. All we had to do was put our ear to the trees and listen very carefully.

Our reconstruction was the first to reach back far enough in time to shed light on the role of the NAO in the single most prominent feature of European climate history: the transition from medieval warmth to the Little Ice Age. It showed that the NAO was mostly positive throughout medieval times but switched to more regular and more negative NAO phases after 1450 (see fig. 10A). What we had found was the driving mechanism behind medieval warmth in Europe: a predominantly positive NAO phase that kept the North Atlantic wind wheel working at full speed and sent warm Atlantic air to central Europe, resulting in the mild winters that allowed Europe's agriculture, culture, and population to flourish. The wheel slowed down and became more erratic after 1450, at which point the cool climate and related hardships of the Little Ice Age started kicking in.

▪ Discovering the mechanism behind European warmth during the Medieval Climate Anomaly was a big deal in our line of business, and we decided to submit our manuscript to *Nature*, one of the top-ranked scientific journals in the world. *Nature* publishes only about 8 percent of the more than 10,000 submissions it receives each year. If the *Nature* editor decides that a manuscript is interesting, she sends it out for peer review. If not, you get a rejection email within about two weeks of submission.

For authors submitting a manuscript to *Nature*, those two weeks are nerve-racking. The NAO manuscript was the first paper I had ever submitted to a top-tier journal. I had a lot riding on it: a yea or nay could make or break my scientific career. I was a postdoc at the time, and a *Nature* publication would go a long way toward securing a professorship. A rejection, on the other hand, would mean that I had spent the past two years on a not very groundbreaking paper. After about 10 days of obsessively checking my inbox, I finally received the email from *Nature*. It was a nay.

From: Patina@Nature.org
To: trouet@wsl.ch
Subject: Nature manuscript 2008-08-08011

Dear Dr. Trouet,

As you know from our previous email we received your manuscript entitled "Pervasive Positive North Atlantic Oscillation Mode Dominated the Medieval Climate Anomaly." Thank you for your interest.

An initial evaluation by our editorial board has found the results to be of broad interest and relevant to many scientific disciplines. The manuscript is well written, graphics of high quality, and represent [*sic*] a step forward in the understanding of the climatic system during the Medieval Climatic Anomaly.

Unfortunately, we cannot consider this research to be novel as we will publish as a Letter in the September 22 Issue a manuscript entitled "The European Medieval Climatic Anomaly was driven by the North Atlantic Oscillation." There appear to be broad similarities in these research questions, results, and implications.

We recognize that some overlap of research efforts is inevitable. This is particularly common within the Medical and Life Science disciplines. While this allows results to be confirmed, refuted and furthers the scientific process, the limited space of *Nature* will not allow us to consider your article further. We congratulate you on your research to date, and wish you luck in publishing your results elsewhere.

Sincerely yours,
Enraldi Patina
Editor, Nature

I had mentally prepared for a rejection, but I was taken aback by the reason for it. Had another research group submitted a similar manuscript? Had we been scooped? Barely having finished reading the email, I stormed into Jan's office to tell him the news. David and Ulf, who from their neighboring offices could hear my outcry, quickly followed. I launched into a speculative rant about potential perpetrators, expecting my colleagues to join in. Instead, they started snickering. As it turns out, in addition to being world-class scientists, my colleagues were world-class pranksters. From the comfort of his office just seconds earlier, David himself had sent me the rejection email from a fake *Nature*-esque address he had invented just for the occasion.* My stunned disbelief that my colleagues would pull such an elaborate prank on me was quickly supplanted by relief at the realization that this was Fake News, that our manuscript had not been rejected yet. Four days later came the real rejection email. The editor had not found our results to be of wide enough interest to warrant publication in *Nature*. But at least we hadn't been scooped.

We decided to rephrase some of the text in our manuscript to clarify its broad impact and submit it to *Science*, another top scientific journal. However, there was a key hurdle to be cleared before we could do so. Jan was convinced that we needed to come up with a catchy title for our paper. He argued for "Wind of Change," which he figured would be a pertinent yet cheeky nod to a famous song by The Scorpions. To be clear: Jan wanted to name our paper, which reported the most important scientific findings of my career to date, after a song by a German eighties rock band that includes the following lyrics:

Take me to the magic of the moment
On a glory night
Where the children of tomorrow dream away (dream away)
In the wind of change.

Needless to say, I put my foot down and we settled on a nice, dry scientific title: "Persistent positive North Atlantic Oscillation mode dominated the Medieval Climate Anomaly." *Science* published the paper about a year later. Maybe not everyone kept reading after the title, but the paper was cited a lot anyway.

---

* He had used @Nature.org rather than @Nature.com.

Eight
Winter Is Coming

The past 1,000 years, the period for which we have the best understanding of climate history thanks to an abundance of tree-ring data and other paleoclimate proxies, is also the best-documented period in human history. As we move closer to the present, the planet becomes more densely populated and societies become increasingly complex. Dense populations and complex societies keep more written records, including trade, naval, and agricultural documents, explicit weather and nature observations, and census data.

Such written records can serve as man-made climate proxies and can contribute to our study of past climate. For instance, starting in the fifth century, with the arrival of Christianity in Ireland, Irish monks faithfully recorded and described notable societal events in the Irish Annals. The more than 1,000-year-long record (431–1649) can be considered a *Game of Thrones avant la lettre* that includes detailed descriptions of wars, political intrigues, and Justinian's plague in the sixth century, but also reports on storms, droughts, and other extreme weather events. Francis Ludlow, a geographer at Trinity College in Dublin, has extracted weather information from the more than 40,000 written entries in the Irish Annals and was able to link particularly cold and severe winters described in the text to past volcanic eruptions. But the use of man-made sources in paleoclimate research is not limited to documents alone. Our ancestors were creative in the ways they commemorated historical climatic events. A 2018 summer drought, for instance, exposed "hunger stones" in the Czech Elbe River. Records of extremely low water levels from the fifteenth through the nineteenth century, as well as warnings of their consequences, have been chiseled into these river boulders. For instance, one carved hunger stone reads, "Wenn Du mich siehst, dann weine." If you see me, weep.

■ The role of historical documents informing us directly about past climate change arguably is overshadowed by their role informing us about its societal impacts. By linking past climate derived from proxies, and human history derived from historical documents, we can examine how people and societies have reacted to climatic changes in the past and, one hopes, learn from history as we move forward in a changing climate. Of course, to examine such connections between history and climate, we need precise dating of both climatic and historical events. With their absolute dating, annual precision, and temporal overlap with documented human history, tree rings are in perhaps the best position among climate proxies to inform us about links between climate history and human history.

A good example is the Mer de Glace (Sea of Ice) glacier, the largest glacier in the French Alps, whose past movements have been recorded in subfossil tree-ring samples that were revealed by a twentieth-century retreat of the glacier. Documentary archives of the town of Chamonix, which was founded in the Mer de Glace valley in the early thirteenth century, tell us about the hefty toll such glacial movements take on local communities. Glaciers retreat when melting, evaporation, and erosion remove more ice from their snout, or terminus, than is transported into it. As the glacier retreats, its terminus moves farther up valley than before, and its moraines, the dirt and rocks that are pushed along by the glacier as it moves, become exposed. With the exposure of moraines, the remains of extensive subfossil forests may become exposed as well. On a recently exposed glacier moraine we might find remnants of trees that started growing in a past period of glacier retreat but were subsequently run over and killed by the glacier as it advanced, only to be revealed again as the glacier retreated in the twentieth century. The wood of such subfossil trees can be well preserved because of the anoxic conditions under the glacier ice, so in many cases it can be tree-ring dated. The date of the outer ring of a glacier subfossil tree can tell us the year the tree was killed by a glacier advance. Since trees can't grow on ice fields, the tree's life span can tell us how long the area was "glacier free" before the advance.

For his *dendrogeomorphological* study—using tree rings to study changes in the Earth's surface over time—of the Mer de Glace glacier, Melaine Le Roy, a PhD student at the University Savoie Mont Blanc in Chambéry, collected such subfossil tree-ring samples on its moraines. Melaine spent several summers surveying sections of the glacier's lateral moraine with a telescope from the opposite side of the valley to locate wood. He then rappelled down from the moraine crest or scrambled up from its base to sample the stone pine (*Pinus cembra*) remains he had observed. Melaine's crossdated Mer de Glace tree-ring samples revealed glacier advances in the late sixteenth and early nineteenth centuries that overran mature forests established in the preceding centuries. Trees started growing on the Mer de Glace glacier moraines during medieval times, when temperatures were warm and the glacier was in a retreated stage. They were then overrun by the advancing glacier during the cool Little Ice Age, before their subfossil remains were found in the twentieth and twenty-first centuries, after the glacier had retreated again.

The Little Ice Age Mer de Glace advances heavily impacted the communities that had settled in the valley, in particular the commune of Chamonix. Chamonix is one of France's oldest ski resorts, and it was the site of the first Winter Olympics in 1924. But before it became a tourist destination and a popular ski town, the parish of Chamonix was not a great place to live. Even during relatively warm medieval times, when Mer de Glace was in a position of retreat, life in Chamonix was hard, though relatively safe. Emmanuel Le Roy Ladurie, a French historian, found Chamonix described in sixteenth-century documents: "The place is among mountains so cold and uninhabitable that there are no attorneys or lawyers to be had . . . there are a lot of poor people, all rustic and ignorant . . . so poor that in the said places of Chamonix and Vallorsine there is no clock by which to see and know the passage of time . . . no stranger will come to live there, and ice and frost are common since the creation of the world."* To me, a place without lawyers, attorneys, or clocks does not sound half bad, but the Chamonix inhabitants lamented their poverty and blamed it on the proximity of the glacier and the brutal climate. And that was before winter came. With a bang.

Starting in the year 1600, in the midst of the Little Ice Age, the Mer de Glace glacier advanced dramatically and in its course erased three hamlets. The Mer de Glace advance was accompanied by avalanches and catastrophic floods

* E. Le Roy Ladurie, *Times of feast, times of famine: A history of climate since the year 1000* (New York: Doubleday, 1971).

that severely damaged the other villages and fields in the Chamonix valley. A commissioner of the Savoie's chamber of accounts who visited one of the hamlets in 1616—just 15 years after the start of the glacier's advance—described the situation as follows: "There are still about six houses, all uninhabited save two, in which live some wretched women and children, though the houses belong to others. Above and adjoining the village, there is a great and horrible glacier of great and incalculable volume which can promise nothing but the destruction of the houses and lands which still remain."[*]

European history is rife with stories of a benign medieval climate giving way to the hardship of a frigid Little Ice Age. According to these stories, 350 years of reliably balmy medieval weather allowed crusaders to recover the Holy Land; architects, masons, and carpenters to erect twenty-six immense Gothic cathedrals in England alone; and Scottish merchants to build castles worthy of noblemen. The warm summers helped more than 50 monastery vineyards to thrive in southern England[†] and were particularly beneficial to the maritime power of the Norse, who ventured as far west as Greenland and Newfoundland. But as the climate shifted in the mid-fifteenth century, those same Viking settlements were deserted, and wine cultivation in southern England was abandoned as summers got cooler and growing seasons got shorter. Advancing glaciers in the Alps and Scandinavia ran over towns and farms, while the Baltic Sea froze over and the northern fishery industry collapsed.

Such stories led early scholars of socioclimatic history to combine climate history and human history in a deterministic approach. They held climate change, and climate change alone, responsible for the ebb and flow of past civilizations. Overall, it certainly appears to be true that the cold Little Ice Age climate created conditions of hardship throughout Europe and the North Atlantic. Risks increased for mountain men and seafarers alike, and food shortages were widespread, resulting in times of starvation, epidemics, social unrest, and violence. But thanks to a growing body of research, we now better understand the complex interactions linking climate history and human history.

---

[*] Communal Archives of Chamonix, CC1, no. 81, year 1616, cited in ibid., 147.

[†] Compared with more than 400 vineyards currently in operation in England and over a much more extensive (northern) area.

As it turns out, the long Little Ice Age winter also had positive outcomes, such as the immortal wintry scenes in Pieter Brueghel the Elder's paintings (ca. 1525–69) and Mary Shelley's *Frankenstein*. Mary Shelley wrote her instant classic while on vacation at Lake Geneva in the summer of 1816. The weather was dreadful that summer, forcing Mary and her husband to stay indoors. To entertain each other, they told horror stories, one of which included a young scientist who created a hideous monster. We now know from early meteorological measurements and tree-ring data that the year of the classic monster's conception, nicknamed the "Year without a Summer," was the result of the eruption of the Tambora volcano in Indonesia the year before. Even the celebrated tones of the Messiah and its Stradivari contemporaries have been attributed to slow tree growth during the cold Little Ice Age. Stradivari's violins (1656–1737) were constructed from spruce and maple trees with very regular, narrow rings and uniform wood density. Legend has it that these narrow rings, resulting from exceptionally cold Little Ice Age summers, contributed to the superior quality of the wood used for the violins.

We have learned that both the winners and losers of the harsh Little Ice Age conditions in Europe were determined as much by differences in societal resilience and adaptation strategies as by regional climatic differences. The Dutch Republic, for instance, experienced its golden age smack in the middle of the climatic turmoil of the seventeenth century. Although the Low Countries were hit just as hard as, if not harder than, the rest of Europe by Little Ice Age frosts, storms, and torrential rains, the Dutch conceived deliberate strategies to profit from the Little Ice Age climate. The Dutch fishery industry flourished when massive schools of herring migrated from the cooling waters of the Baltic Sea toward the North Sea. Dutch farmers created new farming methods and diversified crops, for instance, by including potatoes as a staple. Dutch merchants capitalized on ruined harvests across Europe to raise the price on stockpiled grain and control the European grain supply. Meanwhile, the Dutch Republic invested in its transportation network and welfare programs to absorb the worst impacts of the Little Ice Age chill.

Neither climate history nor human history, nor the relation between the two, however, is linear. The winners and losers of climate change are not set in stone. The history of Norse settlement in Greenland, which turns out to be more complex than first thought, provides a good example of how the resilience and vulnerability of communities can change over time. Throughout medieval times, with Arctic sea ice retreating further and further north as temperatures warmed, the North Atlantic Ocean lay wide open for the Norse to roam ever further westward. Driven by overpopulation and limited options for agriculture in the Scandinavian fjords, they reached and settled the Faroe Islands in the ninth century. From the Faroes, they traveled on to Iceland, where they arrived in 874 and swiftly and completely removed forests from the landscape to make room for agriculture. From Iceland, Erik the Red and his expedition sailed in the late tenth century to Greenland, where they found green summer pastures that could serve as grazing land, as well as abundant fish and sea mammals. Believing that people would be much more tempted to go there if it had an attractive name, Erik called the country Greenland in his reports. His marketing strategies paid off: soon after his return to Iceland, the Norse established two permanent settlements on the southeastern and western coasts of Greenland. From there, they ventured even further west to Newfoundland, where they established a permanent settlement in L'Anse aux Meadows around 1000 CE. For more than two centuries the New World settlements survived and thrived. Greenlanders sailed westward to North America in search of timber and eastward toward Iceland and Norway to pay tithes and trade walrus ivory for supplies.

But the Little Ice Age winter came early in the North Atlantic region, and by 1250 Arctic pack ice* started to reach further south than before, forcing Norse seafarers to alter or even halt their voyages between the Norse colonies and the motherland. As the climate deteriorated, the New World settlements grew increasingly isolated, and farming on Greenland's marginal lands became all but impossible. Glaciers advanced, growing seasons shortened, and supplies from the motherland dwindled. At first, the Greenland Norse adapted well to the worsening conditions by modifying their farming practices, while simultaneously reducing their dependence on agriculture and diversifying their subsistence strategies. In the southeasterly settlement, the colonists developed an irrigation system to boost hay harvests, while colonizers in the western settle-

---

*A nearly continuous mass of floating ice.

ment expanded their hunting grounds in order to obtain more walrus ivory to trade and more seal and caribou for their diet. By the fourteenth century, however, these initial investments in trade and hunting were disrupted as competing Thule communities migrated into Greenland and walrus ivory fell out of fashion in Europe. When a decade of exceptionally cold winters (1345–55) occurred against this less favorable backdrop, it meant the deathblow for the western Greenland settlement. At its peak, the western settlement counted at least 95 farms and around 1,000 inhabitants. But when Ivar Bardarson, a Norwegian cleric, visited the settlement in the late 1350s, all he found were vacant farms. The mid-fourteenth-century abandonment of the western settlement was followed by the abandonment of the eastern settlement a century later.

Yet, in the meantime, the nomadic Inuit, who had cohabited with the Norse in Greenland during medieval times, prospered, rather than failed, during the Little Ice Age. The Inuit traveled and hunted in kayaks, which they launched from the edge of the sea ice and maneuvered through floes, allowing them to hunt year-round. They made use of the expanding sea ice during the Little Ice Age to broaden their hunting "grounds." During the seventeenth century, they were occasionally spotted as far southeast as the Orkney Islands and northern Scotland. Further south on the British Isles, some Londoners also found ways to profit from Little Ice Age conditions. From the seventeenth through the early nineteenth century, winter frost fairs were held periodically on the river Thames. The Thames was wider, shallower, and slower in those days, and in the coldest winters it often froze solid for a few days, allowing horse and coach races to be held and, in one notable instance in 1814, an elephant to be led across the ice. As all Belgians know, it is just a shame that the Brits never made use of the opportunity created by the Little Ice Age decline of their wine industry to learn how to properly brew beer.

Nine
Three Tree-Ring Scientists Walk into a Bar

Tree rings can tell us amazing things, but as climate proxies they have flaws. Trees, in the strict sense of the word, do not grow in oceans or lakes, in the Antarctic, or in vast regions of the Arctic. For lack of distinct annual rings, few trees in the tropics have been used for paleoclimate research. Fortunately, all of these regions are home to other biological and geological archives that can be used to study past climate. Arctic and Antarctic ice can be cored to reveal hundreds of thousands of layers of snow and ice. Ocean and lake sediments also exhibit layers that take us further back in time the deeper we delve beneath the surface. Such layers do not provide the faithful annual record of tree rings; an individual sediment layer can represent five, ten, a hundred, or even a thousand years. Nonetheless, they can inform us about past climates and ecosystems, often over much longer timespans than tree rings can. The oldest core extracted from the Allan Hills in Antarctica, for instance, includes ice that is more than 2.7 million years old, a timespan far beyond the reach of dendrochronology.

However, other proxies, like stalagmites, do form annual layers. Corals, clams, and the ear bones of fish—called *otoliths*—can also reliably form a growth band each year. These "trees of the ocean" can tell us stories about marine currents, temperatures, and atmosphere-ocean interactions, such as ENSO. *Sclerochronology*, a relatively young field of science that borrows many techniques from dendrochronology, including crossdating, uses these marine creatures to develop centuries-long ocean proxy records. Sclerochronologists from Bangor University in Wales, for instance, have developed a 1,357-year-long chronology from the shells of ocean quahogs (*Arctica islandica*), edible clams commercially harvested along the shores of Iceland. Unfortunately, the sclerochronologists may have followed a little too closely in the footsteps of dendrochronologists. Among the quahogs dredged off of Iceland's northern coast was a 507-year-old specimen named Hafrun,* the world's oldest known noncolonial animal. Unaware of Hafrun's advanced age, however, the Bangor University researchers killed it when they cracked open its shell to analyze its growth rings. Like the oldest known living tree, Prometheus, Hafrun was killed for science. Unlike the scarce bristlecone pines, however, quahogs number in the millions in the North Atlantic Ocean. It is unlikely that Hafrun was the oldest one, but searching for another 500-plus-year-old mollusk is like looking for a pine needle in a haystack, and sclerochronologists are in it for the long haul.

■ Bryan Black, a colleague at the LTRR, began his career as a dendrochronologist but then defected to sclerochronology and managed to combine the two fields in his study of California's coastal climate. Bryan crossdated the otoliths in splitnose rockfish (*Sebastes diploproa*) caught along the California coast and used them to develop an almost sixty-year-long chronology of marine productivity. He then compared the rockfish otolith chronology with three time series of Californian seabird† egg-laying dates and fledgling success rates. He found that years with vigorous fish growth corresponded to years with high seabird breeding success, both synchronized by a common driver, the California Current. This ocean current moves southward along the California coast, lifting deep, cold, nutrient-rich waters to the surface, where they support marine ecosystems. The California Current and its upwelling are strong in winters when a high-pressure ridge sits off the Pacific Coast. The clockwise rotation of the winds coming off of such a high-pressure ridge (in the same vein as the clockwise-rotating winds of the Azores High cog) strengthens the southbound California Current and favors upwelling. Such high-pressure systems, despite their positive impacts on marine productivity, simultaneously block North Pacific winter storms from bringing rain and snow to California (fig. 11A). This was the case in the 2012–16 California drought, when a persistent high-pressure ridge was so successful at blocking winter storms from reaching California that it was nicknamed the "Ridiculously Resilient Ridge."‡

---

* Icelandic word meaning "the mystery of the ocean."

† Auklet and murre.

‡ See Daniel Swain, "The Ridiculously Resilient Ridge continues into 2014; California drought intensifies," *The California Weather Blog,* 11 January 2014, http://weatherwest.com/archives/1085.

The 2012–16 California drought is beautifully captured in the tree rings of blue oaks (*Quercus douglasii*) growing in the Central Valley and the foothills of the Sierra Nevada. Native blue oaks are some of the most moisture-sensitive trees to be found anywhere on the planet; not a dry winter has occurred in California over the past 700 years without producing a narrow ring in the blue oak chronology. Blue oak tree rings have therefore been used to study California's hydroclimate history—to reconstruct streamflow in California's major rivers, as well as water quality in the San Francisco Bay. Drought-driven downturns in blue oak productivity can also be linked to the Ridiculously Resilient Ridge and thus back to the California Current. When Bryan compared his time series of rockfish and seabird productivity with the blue oak tree-ring chronology, he found an inverse relationship. In years when the Ridiculously Resilient Ridge blocked winter storms and moisture from reaching the blue oaks in the Central Valley, the California Current was strong and its upwelling supported seabird and rockfish success. While Californians and their blue oaks were suffering through the 2012–16 drought wrought by the Ridiculously Resilient Ridge, their fish and seabird populations were thriving. The moral of Bryan's story? Maybe it's that the ear bones of rockfish and tree rings in California demonstrate more connections than anyone could reasonably imagine. Or maybe the moral is that even if you live in coastal California, you can't have it all.

My team used the same blue oak tree-ring data to study another aspect of the California hydroclimate: the Sierra Nevada snowpack. On April 1, 2015, in the fourth year of the California drought, Governor Jerry Brown announced the first-ever mandatory statewide water restrictions to mitigate its effects. He did so from a bone-dry Phillips snow station just west of Lake Tahoe in the Sierra Nevada, where snowpack has been measured since 1941. Snowpack is often expressed as *Snow Water Equivalent* (SWE), which reflects how much water is stored in snow; it can be thought of as the amount of water the snowpack would create if it melted all at once. SWE is typically measured on April 1, after most of the snow has fallen for the season but before it has started to melt, and April 1 SWE is broadly considered representative of the entire winter season. The average April 1 SWE at the Phillips Station for the years 1941–2014 was 26.8 inches. On April 1, 2015, when Governor Brown made his announcement, there was no snow on the ground. Zero inches.

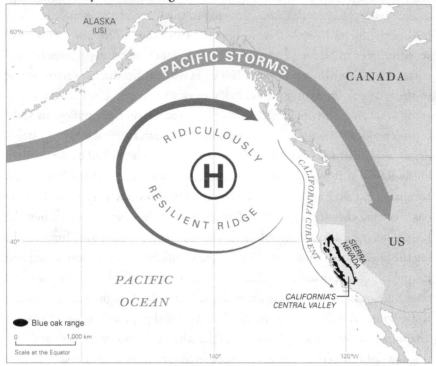

Figure 11A The "Ridiculously Resilient Ridge," which blocks North Pacific storms from bringing rain to blue oaks in California's Central Valley, also blocks them from moving farther east to drop snow in the Sierra Nevada. The clockwise rotation of the winds coming off of the Ridge strengthens the southbound California Current. This favors upwelling and supports marine ecosystems, but it can cause drought in the Central Valley of the Sierra Nevada.

The lack of snow at Phillips station was a yardstick for the entire Sierra Nevada, where the earliest SWE measurements date back to 1930. When the April 1 SWE values for 2015 were released, they showed that the 2015 snowpack was the lowest it had been in more than 80 years. When Soumaya Belmecheri and Flurin Babst—two postdocs in my research group at the LTRR—heard those numbers, they realized that we could use the drought-sensitive blue oak tree-ring data to put the 2015 SWE value in a much longer, perhaps more meaningful context. They were the first to see an important connection: the same Ridiculously Resilient Ridge that blocks North Pacific storms from bringing rainfall to the blue oaks also blocks the storms from moving farther

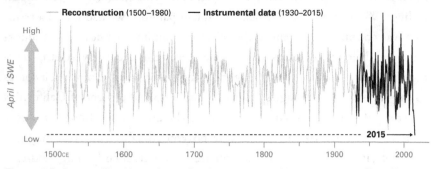

Sierra Nevada Snowpack
1500–2015CE

Reconstruction (1500–1980) ——— Instrumental data (1930–2015)

High

April 1 SWE

Low

1500CE   1600   1700   1800   1900   2000

2015 →

Figure 11B By compiling blue oak tree-ring data, we were able to reconstruct Sierra Nevada snowpack back to 1500. The snowpack reconstruction shows that 2015 was a 500-year low.

east and delivering snow to the Sierra Nevada. In other words, in years when it was dry in the Central Valley, where blue oaks grow and their tree rings record drought, it was also dry in the Sierra Nevada. Through that connection, we could use blue oak tree rings to reconstruct Sierra Nevada annual snowpack levels, not just over the past few decades, but over centuries.

As soon as the April 1 SWE data for 2015 were released, Soumaya and Flurin set to work. They compiled instrumental Sierra Nevada SWE data and more than 1,500 blue oak tree-ring series. They conducted quality checks of the two data sets, calculated calibration statistics, developed the reconstruction model, and then calculated uncertainty intervals, return intervals, and probabilities. The process of developing accurate and reliable climate reconstructions from tree-ring data is heavy on statistics and quantitative analysis. Soumaya and Flurin spent the month of April glued to their computers, coding every piece of the analysis puzzle and double-checking every result. During these long days spent in the glow of screens, I was reluctant to knock on the door of their shared office, worried that I would interrupt their flow. At the same time, I shared their suspicion that we were about to find something big, and I was impatient to see the results.

By early May, after a month of compiling, coding, and debating the best reconstruction approach, we had developed a Sierra Nevada SWE reconstruction that spanned more than 500 years (1500–2015 CE). Our reconstruction showed us that the 2015 Sierra Nevada snowpack wasn't just an 80-year low:

it was a *500-year* low. In the past 500 years, the Sierra Nevada snowpack had never been as meager as it was in 2015 (fig. 11B). This result left us with mixed feelings. We realized that our dedication had paid off; we had found a result that was not only important to the scientific community but also to Californians and their policymakers. At the same time, we all knew very well that finding a 500-year record low is rarely a good thing, and certainly not when it comes to the natural water-storage system responsible for 30 percent of California's water supply. The unprecedented nature of the 2015 snow drought is also a harbinger of things to come: with anthropogenic climate change continuing to accelerate, it is likely that such lows will occur more frequently in the future.

Because of the relevance and urgency of our results, we decided to write up a short paper: 500 words seemed appropriate for a 500-year low.* We aimed to be as swift in our writing as in our analysis. We submitted our paper to *Nature Climate Change* in late May, and our paper was published in mid-September, just five months after the release of the April 1 SWE values. The California drought was still in full force, and our timely message about its extraordinary character drew an unexpected amount of media attention. The *New York Times*, the *Los Angeles Times*, the *Washington Post*, CNN—all of a sudden, everyone was talking about a 500-year record snowpack low. If Soumaya Belmecheri and I thought we were stressed before, we thought again.

Our paper was published on the first day of a two-day project workshop that we had organized in Tucson, to which we had invited a dozen of our collaborators from across the US. In retrospect, we should have asked the journal to postpone publication for a week, but at the time we had no inkling of the level of media interest our paper would receive. On Monday morning I received 25 interview requests in less than 45 minutes. Every radio station in California wanted a soundbite of Soumaya or me talking about snow drought. The workshop was a mess, and we were bad hosts, but we did get the issue a lot of exposure, and the news about the 500-year snow drought spread like wildfire.

In a sense, the Sierra Nevada snowpack plot could be considered the Drought Hockey Stick that I'd been searching for seven years earlier. It was an easy-to-interpret graph that clearly showed the unprecedented nature of the current climate, setting off a media frenzy. Several climate-science deniers lashed out

* Scientific papers typically range from 1,500 to 5,000 words in length.

at us and our results, but their attacks were minor in the grand scheme of things—nothing like the reactions to the inaugural Hockey Stick. Maybe 15 additional years of unabated climate change had made denial harder, or maybe deniers don't care much about snow in California, or maybe they thought Soumaya and I were just two inconsequential girls not worthy of a fight. Regardless, we count ourselves lucky to have largely escaped their wrath.

▪ Our research on the 2015 record low Sierra Nevada snowpack is a good example of how tree rings can be used to study climate extremes. Weather and climate extremes,* such as droughts, heat waves, floods, tornadoes, and hurricanes, are a very destructive aspect of the climate system. These rare, severe deviations from the long-term average climate can threaten, and have devastating effects on, human lives, livelihoods, ecosystems, and the economy. Because extreme events are rare by definition, they are difficult to study. For instance, if you wanted to research category 5 hurricanes in the Atlantic basin—the kind with sustained wind speeds of more than 156 miles per hour—you'd have only 33 events to work with from the almost 170-year timespan since hurricane records started in 1851. Of those 33, only 3 made landfall in the US as category 5 hurricanes (the Labor Day Hurricane in 1935, Hurricane Camille in 1969, and Hurricane Andrew in 1992).† Three landfall events are not numerous enough for us to reliably estimate how frequently they happen, or what their likelihood is of happening in the future.

To improve the reliability of such likelihood estimations, we can use paleoclimate proxies to create longer time series that give us more extreme events to examine. Tree rings are particularly good at capturing extreme events because of their annual resolution. Tree-ring records are most often used to reconstruct droughts and temperature extremes, but they can also be used to reconstruct other climate extremes, such as floods and storms. When a windstorm or hurricane rips off a tree's foliage or breaks off its branches, the canopy damage can be recorded in the tree's rings. When trees lose many of their leaves at once, they lose their capacity to photosynthesize, so they lack the energy to grow

---

* The distinction between the two is not sharp but primarily relates to their timescales. Weather extremes occur on timescales of from 1 day to a few weeks. Climate extremes last at least a month.

† Hurricane Michael made landfall in Florida in October 2018 with wind speeds of 155 mph, just under the category 5 threshold.

wide rings. Without a full canopy and without full photosynthetic capacity, carbon, rather than water availability or temperature, becomes a tree's primary limiting growth factor. Trees that are exposed to storms can record those storms through a period of suppressed growth and a sequence of narrow rings. The growth suppression starts in the year the storm hits and the tree loses its leaves, and it lasts until the tree fully regrows its canopy. There are, of course, other reasons why a tree can lose its photosynthetic capacity (e.g., defoliating insects, fires, competition from other trees) or why its growth can be suppressed (e.g., drought). Therefore, selecting the right site and trees is crucial when studying past storms through tree rings, a research field called *paleotempestology*. A good site for storm reconstruction will have trees that are exposed to storms but are not subject to other limiting factors, such as insect outbreaks or wildfires.

In the Florida Keys, the string of islands off the southern tip of Florida in the northwestern Caribbean, the conditions are just right for paleotempestology. The slash pines (*Pinus elliottii*) that grow on Big Pine Key don't suffer from drought or cold summers, and there are no defoliating insects or other disturbances to speak of. As a Caribbean island, Big Pine Key is frequently hit by hurricanes: since 1851, no fewer than 45 hurricanes (categories 1 through 5) have tracked within a 100-mile radius of the island. Like most of Florida, Big Pine Key is pretty flat. The highest elevation on the island is only six feet above sea level. Still, these six feet are enough to give trees growing in the center of the island an advantage over their lower-elevation neighbors. The trees growing on higher ground are better able to survive the storm surges that come with hurricanes, because the intruding salt water retreats faster, not staying around long enough to kill the trees. Furthermore, the Big Pine Key slash pines have adapted their growth form to the frequent hurricane visits, which makes them resistant to windthrow: they are not easily uprooted or broken by wind. As a result, few slash pines on Big Pine Key die during hurricanes, but many lose leaves and limbs and show related growth suppression in their rings.

When the dendrochronologist Grant Harley, at the University of Idaho, looked at the ring patterns of slash pines he sampled on Big Pine Key, he saw evidence of frequent growth suppressions in the same years in different trees. When Grant tallied up the number of trees with suppressed growth per year, he found evidence of growth suppression in most, or even all, trees on Big Pine

Key for certain years. When he then compared these years with known hurricane events, the reason for the synchronized growth suppressions emerged: Grant's tree-ring record captured 40 of the 44 hurricanes that have tracked close to Big Pine Key since 1851. This connection between growth suppression and hurricanes allowed Grant to use his 300-plus-year-long slash pine chronology to reconstruct hurricane years for Big Pine Key all the way back to 1707.

Grant told me about his Big Pine Key hurricane reconstruction while we were having a drink on the patio of the Hotel Congress in Tucson. It was May 2013, the last evening of the second American Dendrochronology Conference, which had brought roughly 250 dendrochronologists from around the world to the "tree-ring mothership." As fate would have it, the third dendrochronologist at the table that evening was Marta Domínguez-Delmás, a Spanish shipwreck dendroarcheologist at the University of Amsterdam. At the conference, Marta had presented her research on timber derived from shipwrecks from Spain's Age of Discovery. Between Marta's stories of wreck-diving in the Caribbean and Grant's stories of trees stricken by storms on the Keys, we realized that there was a common theme: hurricanes. Before the introduction of steam-powered ships in the nineteenth century, hurricanes were the primary reason for shipwrecks on the transatlantic voyage from Europe to the Americas. They were also the reason why trees on Big Pine Key showed suppressed growth.

Over drinks on the hotel patio, we came up with the idea to combine the Big Pine Key tree rings with Caribbean shipwreck records to reconstruct past hurricanes and perhaps extend the timeline even further back than 300 years. We hypothesized that if hurricanes were the culprit behind past shipwrecks in the Caribbean, then maybe we could use the number of shipwrecks per year as a proxy for hurricane activity. What we needed to test our hypothesis was a shipwreck database that listed the date, location, and cause of past shipwrecks in the Caribbean region. Marta knew just the database we needed: a book by Robert Marx titled *Shipwrecks of the Americas*, a comprehensive collection of roughly 4,000 shipwrecks in the Americas cataloged by year and location. The book was intended primarily for wreck divers and treasure hunters, but it would prove to be a valuable resource for dendrochronologists as well.

When Marta and I set to work tallying up the shipwreck records in Marx's book, Marta persuaded me to work only with the wrecks of Spanish ships. At first I thought she was just being patriotic, but her arguments made sense. The Spanish were the first to make the transatlantic voyage from Europe to the Americas, and from the sixteenth through the eighteenth century their system of treasure-ship convoys, known as the Silver Fleet, was a crucial driver of Spain's economic power and represented a substantial investment by the Spanish government. The details of its voyages, including the number of ships and fleets sent out to the Americas, as well as the number, location, date, and cause of shipwrecks for each year, were therefore meticulously recorded in the Archivo General de Indias (General Archive of the Indies).* The Spanish shipwreck record in Marx's book was based on this archive and the record was long, going back to the late fifteenth century, and well documented. Just what we needed.

Of the many shipwreck entries in Marx's book, we excluded those for wrecks caused by war, pirates, fires, or navigation failures. We then extracted only those wrecks that occurred in the Caribbean region and during hurricane season (July–November). After a couple of weeks of tabulation work, we had in our hands a time series that showed how many ships wrecked per year out of a total of 657 wrecks from 1495, when six caravels were lost in the port of La Isabela, in what is now the Dominican Republic, until 1825. Of course, we know that Marx's book, and hence our database, is not complete. There are shipwrecks lying on the Caribbean ocean floor that have never been recorded, not even in the General Archive of the Indies. We also know that even after our thorough vetting process, it is likely that some shipwrecks in the database were not caused by hurricanes. Our next step, therefore, was to prove that despite its shortcomings, our shipwreck record was a reliable proxy for Caribbean hurricane activity.

The most straightforward way to accomplish this would have been to directly compare shipwreck years with known hurricane years, as Grant did for growth-suppression years in his Big Pine Key tree-ring chronology. But this was not possible, because the shipwreck record ended in 1825, and the instrumental hurricane record only started in 1851—the two time series do not overlap. That was where the Big Pine Key tree-ring record came in. Grant's 300-

---

*The repository of 43,000 volumes and approximately 80 million pages of archival material documenting the history of the Spanish Empire in the Americas and the Philippines (sixteenth through nineteenth centuries), the archive is housed in a dedicated building in Seville, Spain.

year-long hurricane reconstruction (1707–2010) bridged the gap between the instrumental hurricane record (1851–2010) and the shipwreck record (1495–1825). From Grant's earlier analyses, we already knew that years when many trees showed suppressed growth corresponded to hurricane years. Then came our *pièce de résistance*: when we compared the tree-ring record with the shipwreck record, we found that years with many suppressed trees also corresponded to years with many shipwrecks. Even for us, it was astonishing to see the shipwreck years line up so well with the fully independent years of tree-growth suppression.

The only mechanism that can explain why tree growth was suppressed on Big Pine Key in the same years that many ships wrecked in the Caribbean is hurricanes. This result confirmed our initial hypothesis: by combining the two records, we could push our reconstruction of Caribbean hurricane activity back to 1495, the start of the shipwreck record. After the initial excitement of this idea, and of showing that our hypothesis was correct, the project started to dwindle. We had developed a 500-year-long Caribbean hurricane reconstruction but were unsure where to go next. What had we learned from 500 years of reconstructed hurricanes that we had not already known from 150 years of instrumental hurricane data? It took us a while to figure that out.

The answer came to me in a coffee shop in Flagstaff later that summer. I was doing fieldwork in northern Arizona, taking day trips to core trees from home base at a (not exactly lovely) Motel 6. Unfortunately, a few days in, I had contracted a fever and decided to stay in bed. But my resolve did not last long. A cheap motel room during the daytime is a miserable place to be, especially when sick, so I dragged myself over to the closest coffee shop with my laptop to make use of the unexpected pause by having another look at our Caribbean shipwreck work.

I ordered an Americano and found a spot at the window counter, opened my laptop, and pulled up the full-sized plot of our hurricane reconstruction. When my name was called, I picked up my coffee at the counter, and as I walked back to my spot by the window, focused on not spilling my drink, something in the hurricane plot in my near-peripheral vision snagged my attention. From a blurry distance, only one aspect of our hurricane reconstruction stood out: an approximately 70-year dip in the number of shipwrecks in the late seventeenth century (fig. 12). On closer inspection I found that this dip occurred from 1645 to 1715 CE and represented far fewer shipwrecks, and thus hurricanes, than any other period in the previous or following centuries. Once I saw the gap, I couldn't unsee it. I racked my fevered brain to try and remember what else was remarkable about that particular period. When I realized that the hurricane gap overlapped in time perfectly with the Maunder Minimum, the late seventeenth-century period with record low sunspot numbers, I almost shouted *Eureka!* into the void of the coffee shop. This finding definitely brightened my mood and instantly cured my flu, but none of the other customers in the coffee shop seemed very impressed by my newfound jauntiness. They just kept staring at their own computer screens, muffled in the headphones they had donned to tune out the emo-rock emerging from the coffee shop's speakers. I had to wait until evening to share my excitement with my colleagues coming back from the field.

Shipwrecks, Hurricanes, and Sunspots
1500–2000CE

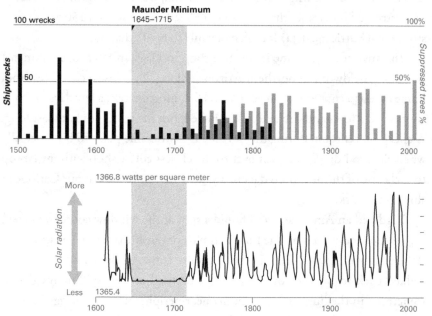

Figure 12 The period from 1645 to 1715 shows a sharp drop in the number of hurricanes and shipwrecks. This gap aligns with the Maunder Minimum, when sunspot numbers were at a record low.

Our hurricane gap was a new piece in the global Maunder Minimum and Little Ice Age climate puzzle. We put the Caribbean on the map of regions where the Maunder Minimum had had an important climatic impact. We had found a long-term link between low solar radiation and low hurricane activity that can be explained by ocean temperatures. Warmer ocean waters provide more thermal and kinetic energy and increase the possibility of hurricane formation. Hurricanes need ocean temperatures of at least 82°F to form and prosper. That is the main reason why hurricane season in the Caribbean typically lasts through the warmer months of July through November. During the Maunder Minimum, the amount of solar energy that reached Earth's surface was lower than normal, leading to cool temperatures in the Atlantic Ocean and the Caribbean. With ocean temperatures cooler than normal, there were fewer opportunities for hurricanes to form. As a result, hurricane activity, and with it the number of shipwrecks, was at a lull during the Maunder Minimum. This connection not only enhances our understanding of the climate of the past but may also help improve the critical climate models that predict what will happen to Caribbean hurricanes in the future, so that we can be optimally prepared for what is to come.

Global climate models (also called general circulation models, or GCMs) are computer-run programs that use the laws of physics, fluid motion, and chemistry to mimic our complex climate system in order to study its dynamics. The more than thirty climate models that are commonly used generally project that with future climate change, tropical cyclones (including hurricanes in the Caribbean) will become less frequent but more intense. Different climate models agree remarkably well on this conclusion for Earth as a whole, but for individual ocean basins there are many discrepancies between models and thus greater uncertainties about what the future will bring. The largest discrepancies for the North Atlantic Basin result from our limited understanding of how hurricanes respond to changes in Earth's radiation budget, which equals the difference between the amount of energy Earth receives from the sun and the amount it emits and reflects back into space. In the twenty-first century, the man-made enhanced greenhouse effect is the main driver of such changes in the radiation budget, but in the past they have also been related to natural changes in solar radiation. The most dramatic of such solar-radiation changes in recent history is the Maunder Minimum. The 75 percent reduction in hurricane activity we found for the Maunder Minimum can now be used as a benchmark for future climate-change modeling studies. Models that show reduced hurricane activity during the Maunder Minimum when hindcasting are likely more reliable than models that don't show the reduction.* Knowing this, we can weight future model forecasts toward the more reliable models.

Our findings can also be used to better understand the potential effect of the Maunder Minimum and its cooling on human history. It is possible that the lull in hurricane activity during the Maunder Minimum influenced European history through its impact on the Spanish transatlantic trade and thus on the economic and political power balance in Europe. When Grant talked about our study in a seminar at the University of Southern Mississippi, he drew the attention of a colleague who specialized in the geography and history of piracy. If you're starting to think that being a dendrochronologist sounds like it could be a cool job, especially if it involves diving for shipwrecks, then please consider that there are people out there who study pirates for a living.

As it turns out, the hurricane-poor Maunder Minimum coincides with the period historians refer to as the Golden Age of Piracy. This period from ca. 1650 to 1720 was the heyday of Anglo-French buccaneers and privateers in the Caribbean, who attacked Spanish ships on their return voyage to Europe. Notorious pirates such as Edward "Blackbeard" Teach, "Black Sam" Belamy (the Robin Hood of pirates), and Anne Bonny (who partnered with Mary Read and John "Calico Jack" Rackham) raided ships that were loaded with gold, gems, and sugar on their way back to Spain. The lack of hurricanes and shipwrecks during the Maunder Minimum would have upped the bounty to be had for the buccaneers and boosted the incentive for pirates to pillage and plunder. And while fewer storms led to fewer shipwrecks on the Spanish side, calmer waters also benefited the pirates, allowing their fortunes to flourish. Of course, geopolitical factors, such as international relations on the European continent and the lack of effective government in the American colonies, played key roles in the seventeenth-century rise of piracy, but it is likely that Blackbeard and Anne Bonny thrived owing to a literal lack of *shiver in their timbers*.

* Climate models are run forward in time from the present to *forecast* how Earth's climate will respond to rising concentrations of atmospheric greenhouse gases in the future. But they can also be started at some point in the past—for instance, at 1000 CE—and run to the present to *hindcast* the past climate.

Ten

Ghosts, Orphans, and Extraterrestrials

Dendrochronology can be used to study not only past climate and its extremes, such as hurricanes, but also other natural hazards. Earthquakes, for instance, can damage trees and affect their growth. The surface rupture, displacements, elevation change, and shaking that occur along tectonic-plate boundaries as they lock, strain, and release during an earthquake can result in tree-ring anomalies that can be used to date and study past earthquakes. While earthquakes are most commonly measured on the Richter scale, which classifies the magnitude of earthquakes based on the strength of their seismic waves, they can also be classified based on the severity of their impact. The Modified Mercalli scale,* which measures earthquake intensity on a scale from 1 through 8, is partly based on the degree of tree disturbance caused by an earthquake. An earthquake is classified as level 5 if trees are shaken "slightly" but level 8 if they are shaken "strongly." Such shaking and related damages can disrupt the normal growth in affected trees and leave a permanent mark in their tree-ring record.

Sometimes, however, the earthquake damage to trees is so severe that the trees do not survive. This is often the case for trees growing near the earthquake's epicenter, but it can also happen when earthquakes in coastal regions create land subsidence and saltwater surges. When an already low-lying area of coastal land sinks an additional couple of meters as the result of an earthquake, it can be submerged by seawater surges that bring sand and silt. The saltwater surge kills trees in its path, and in the decade following the earthquake the former landscape, including its meadows, thickets, and trees, is buried by a layer of sand and silt. The silt layer is often depleted of oxygen, creating anoxic conditions in which the remains of the killed trees (snags) are preserved and

---

*This is a modified and improved version of the original Mercalli intensity scale, developed in 1902.

leave a lasting record of the earthquake's impact. In the case of most trees fallen victim to earthquakes, only the stumps beneath the surface are preserved. Some tree species, however, have wood that is so rot resistant that their aboveground portions also persevere through time after the trees die. Western red cedars (*Thuja plicata*), for instance, can remain standing upright in the marsh for centuries after an earthquake. In a *New Yorker* piece from 2015, Kathryn Schulz describes such trees in *ghost forests*: "Leafless, branchless, backless, they are reduced to their trunks and worn to a smooth silver-gray, as if they had always carried their own tombstones inside them."[*]

Brian Atwater and David Yamaguchi were working for the US Geological Survey when they sampled four such ghost forests in southern Washington, on the North American Pacific Coast, and crossdated their western red cedar snags. They dated the outer ring in every ghost tree they sampled to 1699.[†] Atwater and Yamaguchi found no sign of a slowdown in tree growth prior to the ghost trees' decease, suggesting that the trees were healthy and happy right up to their sudden death. These ghost trees clearly did not fade away; they burned out in a single, abrupt event. The researchers hypothesized that a large earthquake in the winter of 1699–1700, followed by subsidence and a saltwater surge, caused their untimely demise. Their hypothesis was confirmed by nearby trees on higher ground that had survived the earthquake but still showed almost a decade of suppressed growth starting in 1700. Further evidence for the exact calendar date and the magnitude of the 1700 earthquake came from an unexpected source.

The biogeography of the Pacific Northwest, where the evergreen forests of the Cascade Mountains and the Columbia River meet the mighty Pacific Ocean, has long inspired natural scientists, environmental activists, and writers alike. Yet, there are no written records describing the Cascadian bioregion prior to the eighteenth century, when early European explorers, such as James Cook in

---

[*] See "The really big one," *New Yorker*, 13 July 2015, https://www.newyorker.com/magazine/2015/07/20/the-really-big-one.

[†] If we are to date the year of death of a tree, the tree's final ring must be preserved, which was not the case with the weather-eroded trunks of the ghost trees. Atwater and Yamaguchi therefore unearthed the roots of the ghost trees that had been submerged in the marsh and still bore bark; the final ring under the bark was present in these roots. Atwater and Yamaguchi then crossdated the root chronologies to the previously developed stem chronologies to determine the date of the final ring.

the 1770s, followed by Lewis and Clark in the early 1800s, reached the North Pacific Coast. The Chinook and Sahaptin language groups of the Pacific Northwest do have oral traditions that include stories of coastal flooding and subsidence. Without a written record, however, many of these stories have been lost over time, and the ones that have survived do not include specific dates or places. This means that the 1700 earthquake predates the earliest written accounts in the Pacific Northwest by almost a century.

Five thousand miles away, on the other side of the Pacific Ocean, however, the Japanese have been keeping written records since at least the sixth century. As it happens, the 1700 earthquake occurred in the midst of the largely peaceful Edo shogunate (1603–1867), which had little need for a strong military class, so many Edo samurai swordsmen were employed as scribes. Literacy became widespread in Japan during this period, with merchants and peasants contributing to literature and bureaucracy alike. When on January 27 and 28, 1700, a tsunami ravaged more than 600 miles of Japan's Pacific Coast, it was recorded in hundreds of accounts of shipwrecks, floodings, and drowned agricultural fields. The 1700 tsunami is one of the best-documented events in Japan's history, but there was not a single report of an earthquake in Japan that might have caused it. For almost 300 years, the origin of the 1700 tsunami remained unknown. Earthquake historians called it an "orphan tsunami."

In 1997, Atwater and Yamaguchi collaborated with Japanese earthquake historians to reunite Japan's orphan tsunami with its Cascadian parent earthquake. The research team ran computer simulations indicating that the 1700 Cascadian earthquake recorded in the ghost tree rings generated a tsunami front that probably needed about ten hours to reach the northeastern coast of Japan on January 27. The parent earthquake must therefore have happened on the evening of January 26, 1700, and it must have been giant—at least a magnitude 9 on the Richter scale—to have flooded such an extensive area of land 5,000 miles away. An earthquake of anywhere near this magnitude has never occurred in the Pacific Northwest since, so these findings are our most visceral warning of the risk of large, destructive earthquakes in the region. We know from geomorphological data that over the past 3,500 years, large earthquakes have occurred in the Pacific Northwest on average every 500 years. The intervals between earthquakes, however, range from a few centuries to a millennium. The inevitability of the next destructive earthquake in the region is now

Ghosts, Orphans, and Extraterrestrials

accepted, but we cannot yet predict whether it will happen 1 year or 1,000 years from now. What we can do is properly prepare cities, buildings, communities, and people for a potential magnitude 9 earthquake and possibly a related tsunami. Thanks to ghost-forest dendrochronology and the discovery of the 1700 earthquake, public-safety efforts, including tsunami warnings and seismic hazard maps, are now in place that will aid in making the impact of the next event less destructive.

▪ Severe extreme events such as earthquakes can suppress the growth of trees for years and generate long sequences of narrow rings. But some extreme events are so sudden and disruptive to the tree's growth that the core structure of the year's ring, its wood anatomy, is impacted. This impact is particularly significant if it happens during the growing season, when the tree is actively forming new wood cells in a new ring. The anatomical anomalies created by such events stand out in a tree-ring sequence, creating a lasting record of something having gone seriously awry in a particular year.

The nuclear fallout from the disastrous accident at the Chernobyl power plant in Ukraine on April 26, 1986, killed all trees in a 1.5-mile radius, creating the Red Forest, named for the burnt-sienna color of the dead pines. The Red Forest was bulldozed and buried under a thick carpet of sand during the postdisaster cleanup, and its site remains exceedingly contaminated. Trees further away from the power station also suffered severe radiation damage; however, they have been left in situ, providing a unique opportunity to study the effects of radiation on tree growth in the field. Timothy Mousseau, of the University of South Carolina, and his colleagues waited until 2009, 23 years after the accident, to enter the 19-mile core of the military-controlled Chernobyl Exclusion Zone to collect tree cookies for their study. Even after 23 years, they had to wear radiation-protection suits to sample in the most contaminated areas. They collected cross sections of more than 100 Scots pines (*Pinus sylvestris*) and found high concentrations of radionuclides—radioactive atoms—in the wood of all the trees they sampled. Trees growing in the zone absorb high doses of radionuclides through their roots and incorporate them in their wood as they grow. Mousseau and his team found that the closer to the power plant the trees grew, the higher the radionuclide concentrations. Clearing these radioactive forests is too dangerous and expensive to be realistic, but if the pine trees die because of wildfires, droughts, or insect attacks, then the radionuclides will be released into the air and the radioactive particles can be carried for great distances across the Eurasian continent. For now, the risk of this happening is mitigated by teams of Ukrainian firefighters who spend their summer days on the lookout for wildfires from rusty watchtowers high over the Chernobyl landscape.

In addition to radionuclides, Mousseau and his collaborators also found that the nuclear fallout resulted in severe growth suppressions in the rings following the 1986 ring. The 1987–89 rings were most strongly suppressed, but the suppression effect lingered for up to two decades. They also found abnormalities in the wood anatomy of the 1986 ring (fig. 13). In normal pinewood, the cells are organized in straight, undivided rows perpendicular to the tree-ring boundary. In the 1986 ring of the Chernobyl pines, some cell rows converged, while others divided into multiple rows. Yet other cell rows divided first and then later merged again. All of these anomalies indicate radiation damage to the cambium of the tree, where new wood cells are formed. The closer the trees were located to the power station and the accident, the more frequently these oddities appeared.

Dendrochronologists discovered similar growth suppressions and anatomical abnormalities in the trees growing near Tunguska, in eastern Siberia, in 1908. Around 7:00 a.m. on June 30 of that year, a meteoroid entered the stratosphere and disintegrated three to six miles above Tunguska. The meteoroid did not hit the surface of the earth and did not leave a crater, but its entry into our atmosphere created an explosion with a shock wave that would have measured 5.0 on the Richter scale if that scale had existed then. There are eyewitness accounts of the event from Tungusic people as well as Russian settlers who lived close enough to see a fireball moving across the sky, to hear its noise as loud as cannon fire, and to feel the earth shake on its entry. According to the *Sibir* newspaper for July 2, 1908, "In the north Karelinski village the peasants saw to the northwest, rather high above the horizon, some strangely bright bluish-white heavenly body, which for 10 minutes moved downwards. . . . As the body neared the ground, the bright body seemed to smudge, and then turned into a giant billow of black smoke, and a loud knocking was heard as if large stones were falling, or artillery was fired. All buildings shook. At the same time the cloud began emitting flames of uncertain shapes. All villagers

Radiation Damage

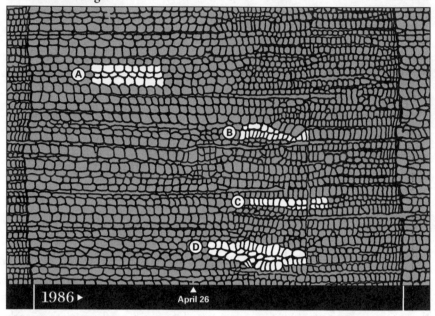

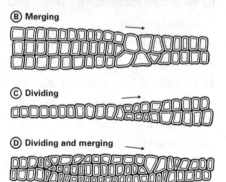

Figure 13 Fallout from the accident at the Chernobyl nuclear power plant in late April 1986 caused severe radiation damage in surviving pine trees. In normal pinewood, the cells are organized in straight, undivided rows perpendicular to the tree-ring boundary (A). In the 1986 ring of Chernobyl pines, some cell rows merged (B), while others divided into multiple rows (C). Yet other cell rows divided first, to then later merge again (D).

were stricken with panic and took to the streets, women cried, thinking it was the end of the world."

Fortunately, Tunguska is in the middle of nowhere; no one actually lived close enough to get a better look, and there were no known human casualties. The first expedition that set out to investigate the Tunguska event reached the site in 1927, almost two decades after the explosion. The explorers did not find a crater, but they did find a 5-mile-diameter ground zero,* where all the trees were scorched, debranched, and killed, though still standing upright. Trees outside this core area were partly burned and had toppled away from the epicenter. From later aerial photographs the magnitude of the impact emerged: the explosion uprooted an estimated 80 million trees in a giant butterfly pattern covering approximately 800 square miles of Siberian taiga.†

Few trees in the Tunguska area survived the blast, but those that did carry the memory of its impact in their 1908 ring. In 1990, the Russian dendrochronologist Evgenii Vaganov, of the Sukachev Institute of Forest in Krasnoyarsk, cored 12 trees that had survived the explosion within a 3- to 4-mile radius of the Tunguska epicenter. Evgenii was curious about the impact of various mechanical aspects of the blast wave—scorching, defoliation, shaking—on the woody growth of the trees. He found effects on the Tunguska trees similar to those Timothy Mousseau discovered in the wake of the 1986 Chernobyl accident: suppressed growth for four to five years following the explosion, as well as severe wood anatomical abnormalities in the 1908 ring itself. The nature of the wood anatomical anomalies, however, was different from that of anomalies caused by radioactive radiation in Chernobyl. The larch, spruce, and pine trees in Tunguska formed *light rings*, rings that have latewood cells with smaller than normal diameters and nonthickened cell walls, in 1908. The trees also formed fewer latewood cells in 1908 than in other years, and the combination of few latewood cells with thin cell walls gives the 1908 ring an abnormally bleached appearance. Such light rings were likely caused by the explosive blast ripping the needles off the tree. Defoliation in the middle of the growing season can stop growth-hormone supplies from reaching the cambium. Without this hormone-driven cambial activity, the tree will run out of energy and the incentive to form new wood or even to properly finish the wood cells that it started forming prior to defoliation.

■ Events do not need to be as dramatic as a nuclear-power-plant meltdown or a meteoroid explosion to leave a lasting mark in the wood anatomy of tree rings. Extreme weather events, such as floods or frosts, can also cause ring ab-

*The point on the earth's surface directly below an explosion.
† The butterfly pattern had a "wingspan" of 43 miles and a "body length" of 34 miles.

normalities. *Flood rings* occur in trees growing on riverbanks, where they may be inundated by spring or summer floods. In severe floods, such inundations can last a long time and create anoxic conditions for the roots and stems of the trees. Such anoxic conditions can lead to growth-hormone imbalances in riparian trees and abnormalities in the wood anatomy of the rings that are formed during the inundation. In oaks, for instance, flood rings have much smaller earlywood vessels—the specialized water-conducting cells in the wood of broadleaf trees—than normal rings. By analyzing the flood rings in oak trees from a wet bottomland forest at the southeastern tip of Missouri, for instance, Matt Therrell and Emma Bialecki, of the University of Alabama, reconstructed Mississippi floods back to 1770 and added 17 spring floods to the instrumental Mississippi record. Their research has shown that twentieth-century engineering modifications, such as the channelization following the great flood of 1927, have elevated the current flood hazard in the Mississippi River system to unprecedented levels. Well-intended engineering efforts to tame the river and confine its wiggle room have only made its flooding worse. Similar warnings about flood-control measures actually increasing flood risk date as far back as the 1850s, long before the Mississippi River channelization even started. But even though the evidence for this is now also clear in the tree rings, so far it has not led to any policy changes.

Ring abnormalities can also occur when frost, with temperatures falling below freezing, occurs during a tree's growing season, when it is forming wood cells and thickening their walls. The dehydrating effect of the frost can severely injure the cambium, which then produces irregularly shaped wood cells. In *frost rings*, a band of deformed wood cells is visible between the neat rows of cells prior to and following the frost event (fig. 14). This band of deformed cells can occur in the earlywood, when caused by a late spring frost, or in the latewood, when caused by an early fall frost.

Dendrochronologists have discovered that sometimes frost events are unrelated to winter arriving early or outstaying its welcome. An unexpected link between frost rings and past volcanic eruptions was first discovered in the early 1980s by the LTRR's Val Lamarche and Katie Hirschboeck. They found frost rings in the then 4,000-year-long bristlecone pine tree-ring record that coincided in trees growing in California and Colorado. Bristlecone pines growing more than 800 miles apart formed frost rings in the middle of sum-

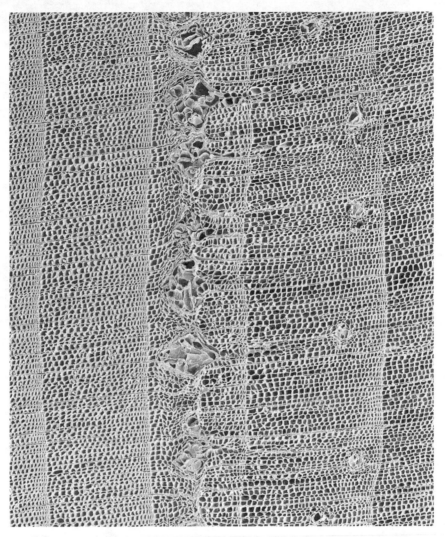

Figure 14 This tree-ring sequence from a Siberian pine (*Pinus sibirica*) in Mongolia covers the period 534–39 CE and includes a frost ring in 536. The narrow ring in 537 also implies unusually cold weather. The cold summers were caused by a volcanic eruption in 536 that kicked off the Late Antique Little Ice Age. Photo by Dee Breger.

mer in the same years, for instance in 1817, 1912, and 1965 CE. The list of frost-ring years rang a bell for Lamarche and Hirschboeck, because the majority of these years closely followed well-known volcanic eruptions. For instance, the 1817 frost ring followed the 1815 Tambora eruption in Indonesia, the 1912 frost ring corresponds to the 1912 Katmai eruption in Alaska, and the

1965 frost ring followed the 1963 Agung eruption in Bali. The two or three years of widespread cooling following volcanic events can thus explain the occurrence of out-of-season but synchronized frost rings.

We have learned a lot about the link between tree rings and volcanic eruptions in the thirty years since Lamarche and Hirschboeck's groundbreaking work. We now know that postvolcanic cooling is recorded not only as frost rings in some trees but also as an extremely narrow ring in most temperature-sensitive trees. The stratospheric aerosol veil produced by major volcanic eruptions can cool temperatures over large regions of the earth's surface for up to two years or more. Tropical eruptions in particular can leave a vast, cooling fingerprint on temperatures and temperature-sensitive tree-ring chronologies worldwide. When we develop hemispheric- or global-scale temperature reconstructions, such as the Hockey Stick or the Spaghetti Plate, we average a multitude of such temperature-sensitive tree-ring chronologies from a range of locations. Years of widespread warming or cooling stand out in such reconstructions because they are common to the majority of contributing tree-ring chronologies. In the Hockey Stick curve, the warm 1990s (the "blade") stood out because most tree-ring chronologies recorded unprecedented warming in this decade. On the opposite side of the scale, the "Year Without a Summer," in which Mary Shelley wrote *Frankenstein*, following the 1815 Tambora eruption, stands out as a cold year in many tree-ring-based temperature reconstructions because many series show an exceptionally narrow ring.

Tree-ring-based temperature reconstructions can therefore be used to study and quantify the effect of past volcanic eruptions on climate change. However, to correctly and confidently attribute past climatic cooling to past volcanic events, we also need independent and precisely dated records of past volcanic eruptions. Such volcanic proxies are provided by ice-core records from Greenland and Antarctica. The sulfuric aerosols that are released when large volcanoes erupt are captured in the snow and ice layers of these ice fields as sulfate (SO_4^{2-}) deposits, and the volcano-derived sulfate spikes can be dated with the ice-core layer in which they were deposited.

But we can only line up the volcanic events recorded in ice-core layers with the cooling recorded in tree rings if both records are absolutely and reliably dated. Crossdating guarantees this for tree-ring records, but ice-core records are prone to more chronological errors. When the paleoclimatologists Michael Sigl, Joe McConnell, and their team from the Desert Research Institute, in Reno, Nevada, compared sulfate spikes in five ice-core records from Antarctica and Greenland with cold years in tree-ring-based temperature reconstructions* for the Northern Hemisphere, they found that the ice-core and tree-ring records lined up perfectly back to 1250 CE. Every large volcanic eruption captured in the ice cores was followed one to two years later by strong cooling in the tree-ring records. However, the 1257 Samalas eruption in Indonesia was the earliest volcanic event that matched up with subsequent narrow rings and cooling, in 1258. Prior to 1250, volcanic eruptions recorded in the ice cores were followed by cooling recorded in tree rings not one or two years but seven years later (fig. 15). Because tree-ring data are crossdated, it is virtually impossible for the tree-ring record to have been misdated by seven years. The seven-year discrepancy before 1250 between volcanic eruptions and subsequent cooling suggests a chronological error in the earliest part of the ice-core record.

The riddle of the pre-1250 seven-year discrepancy was solved in 2012, when Fusa Miyake and colleagues from the Solar-Terrestrial Environment Laboratory at the University of Nagoya in Japan discovered a radiocarbon (carbon-14, or C14) spike in the tree ring from the year 775 CE. Prior to their study, radiocarbon had only been measured in groups of 10 or more consecutive tree rings to contribute to the calibration of the international radiocarbon calibration curve, used to derive an object's calendar age from its radiocarbon content. Averaging C14 values over multiple years is appropriate for this purpose, but it masks peaks in the C14 record that occur in individual rings.

To study such annual C14 peaks and their origin, Miyake and his team measured C14 in individual rings, rather than in 10-year chunks, of two Japanese cedar (*Cryptomeria japonica*) trees. They found a C14 peak in the 775 CE ring of both trees that was about 20 times higher than the long-term C14 average. This 775 radiocarbon peak spans both hemispheres and has since also been found in trees from Germany, Russia, North America, and New Zealand. Only a very strong solar flare,[†] during which the sun ejects massive amounts of radiation in Earth's direction, could have caused atmospheric C14 to change that drastically, that suddenly. The *superflare* that created the 775 C14 peak

---

*They used five temperature reconstructions from central Europe, Scandinavia, Siberia, and the western US.

† Also called a solar proton event, or SPE.

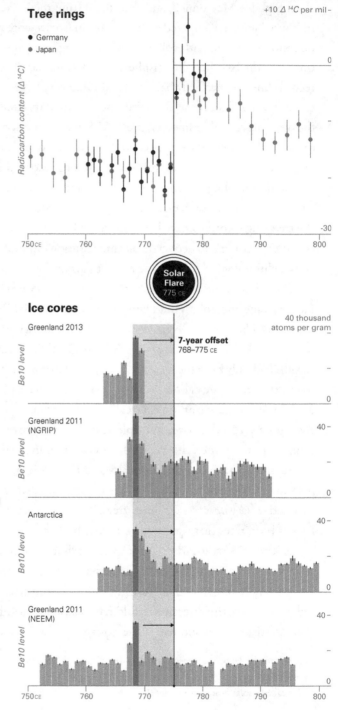

Tree rings

+10 Δ ^{14}C per mil

● Germany
● Japan

Radiocarbon content (Δ ^{14}C)

0

-30

750 CE   760   770   780   790   800

Solar Flare 775 CE

Solar Superflare Recorded in Trees and Ice

750–800 CE

Ice cores

Greenland 2013

40 thousand atoms per gram

Be10 level

7-year offset
768–775 CE

0

Greenland 2011 (NGRIP)

40

Be10 level

0

Figure 15 The 774–75 solar flare was recorded as a radiocarbon spike in the 775 ring of trees across the world. In ice cores, it was recorded as a spike in beryllium-10 levels. However, this spike in ice-core records was recorded in the 768 layer, revealing a 7-year dating error in the pre-1250 part of the ice-core record.

Antarctica

40

Be10 level

0

Greenland 2011 (NEEM)

40

Be10 level

0

750 CE   760   770   780   790   800

must have been 40 to 50 times larger than any that have been directly observed. During a superflare, the sun produces blazes of intense radiation that create unusually high numbers of cosmogenic isotopes, such as C14, when they reach the earth's atmosphere. These cosmogenic isotopes are absorbed by trees through photosynthesis and recorded in the tree rings as C14 peaks. The 775 peak in C14, recorded in tree rings around the world, was likely caused by a superflare in the preceding year, 774. A similar but weaker solar flare created a second C14 peak in tree rings formed in 994.

The sun constantly produces radiation that creates cosmogenic isotopes in the earth's atmosphere. However, solar flares strong enough to result in C14 peaks such as seen in the 775 and 994 tree rings are rare, and the 774 superflare is estimated to have been the strongest over the past 11,000 years. The 774 event was recorded in the eighth-century section of the Anglo-Saxon Chronicle* as follows: "Annus Domini 774. . . . This year also appeared in the heavens a red crucifix, after sunset; the Mercians and the men of Kent fought at Otford; and wonderful serpents were seen in the land of the South-Saxons." Even though this particular chronicler seems to have been charmed by red crucifix auroras and wonderful serpents, the great infrequency of such massive solar storms is a good thing; they potentially deplete Earth's ozone layer, disrupt its geomagnetic field, and seriously mess with our technology and telecommunication systems.

In addition to radiocarbon, beryllium-10 (Be10) is formed in the atmosphere when it is bombarded with the sun's rays. Unlike C14, which is not typically captured in ice cores,[†] atmospheric Be10 is deposited in the snow and ice layers of Greenland and Antarctica. In the same way that sulfate peaks in ice cores can be used as a proxy for volcanic activity, Be10 peaks can be used as a proxy for the sun's activity. Solar superflares, such as the ones in 774–75 and 994, create Be10 spikes in ice cores, which can be used to directly link the date of an ice layer to the date of a related C14 peak in the tree-ring record.

When Michael Sigl and his colleagues measured Be10 concentrations in ice cores from both Greenland and Antarctica, they found a Be10 peak in 768

---

[*] A collection of Old English annals chronicling the history of the Anglo-Saxons from 60 BCE to 1116 CE. Nine copies of the Chronicle survive, seven of which reside in the British Library in London.

[†] Because ice—unlike trees—does not photosynthesize.

and another peak in 987. Both peaks occurred exactly seven years prior to the tree-ring C14 peaks of 775 and 994, suggesting a seven-year offset in this early section of the ice-core record (see fig. 15). This seven-year offset explains the seven-year discrepancy between early volcanic eruptions in the ice-core record and cool years in the tree-ring record. Sure enough, when Michael and his team shifted the pre-1257 section of the ice-core chronology seven years toward the present, all of a sudden the ice-core and tree-ring data lined up perfectly. Fifteen of the sixteen coldest years in the tree-ring record for the period 500 BCE to 1000 CE now followed large volcanic sulfate peaks in the ice-core record by 1 to 2 years. This is how Joe McConnell describes the breakthrough in a *Los Angeles Times* interview: "Before this work, the tree rings and the ice core records diverged. In the new dating, they line right up. We can look at the tree and say, 'There's the cooling that's associated with this volcanic event.'"[*]

Why was a mainstream media outlet like the *Los Angeles Times* interested in interviewing Joe? Because of the importance of this finding for exploring the links between climate and civilization. With volcanic eruptions now securely dated over the past 2,500 years, their impact on climate and human history can be examined. Volcanic eruptions not only cool the earth's surface, they can also impact regional hydroclimate. Nile River failures, for instance, have been linked to volcanic activity. Ample historical records dating back to the time of the pharaohs describe the Nile bursting over its banks every summer, flooding the adjacent plains. When the floodwaters receded in early fall, they left behind fertile black silt that enabled the agriculture that sustained the Egyptian people in an otherwise arid environment. Egyptian societies used *nilometers* to record the rise and fall of their life-bringing river. At Rhoda Island, in central Cairo, the nilometer is a vertical column submerged in the river, with markings at intervals to indicate the depth of the water. Nilometer measurements at Rhoda Island were first taken after the Arab invasion in 622 CE and continued until 1902, when the construction of the first Aswan Dam made them obsolete.[†] Thus the Cairo nilometer provided one of the longest, near-continuous documentary hydroclimatic time series in existence. When

[*] Eryn Brown, "Ice cores yield history of volcanic eruptions, climate effects," *Los Angeles Times,* 10 July 2015, http://www.latimes.com/science/sciencenow/la-sci-sn-volcanoes-climate-history -20150710-story.html.
[†] The dams first reduced and then eliminated the Nile's annual inundation.

the historian Joe Manning, of Yale University, in collaboration with Michael Sigl and others, compared the nilometer record to the now correctly dated ice-core-based volcano record, he discovered that Nile summer flooding was on average almost 9 inches lower in years of large, explosive volcanic eruptions than in other years. Because of the Nile's critical role in agriculture, such low water levels often led to famine.

The researchers then extended this relationship between volcanic activity and Nile failures further back in time to Ptolemaic Egypt. The Ptolemaic Kingdom (305–30 BCE) was founded by a Greek royal family after Alexander the Great's death in 323 BCE, but the Hellenistic rulers frequently had to fight off revolts of the native Egyptian population. When Manning compared the onset dates of these Egyptian revolts, as recorded in papyri and inscriptions, with the ice-core-based volcanic-eruption chronology, he found that an above-average number of revolts started in years of volcanic eruptions or in years immediately following them. This result implies that volcano-induced Nile failure might have been a catalyst for revolt in this agricultural society. For instance, the 20-year Theban revolt that started in 207 BCE followed two years after a 209 BCE volcanic eruption in Iceland.*

Increased instability at home because of the revolts might then have restricted the Ptolemaic rulers in their warfare with neighboring states. Ptolemaic armies were recalled to Egypt to suppress domestic unrest, and dynastic revenues were redistributed from military campaigns to relief efforts to deal with the hydrological shock of Nile failure. In addition to more revolt-onset dates, Manning and his team also found more end-of-war dates corresponding to volcanic years than to other years. Also, priestly decrees, such as the one issued in Memphis, Egypt, in 196 BCE and recorded in three languages on the Rosetta Stone, were issued more frequently in eruption years than in other years. It appears that one aim of such decrees was to reinforce state authority, for which there was more need in times of social unrest, by linking it to the influential priestly class.

---

* The same volcanic eruption also caused famine in China. A first-century BCE Chinese record states that in November of 207 BCE "the harvest had failed, the people were destitute, and for lack of grain [even] soldiers were eating beans and taros." K. D. Pang, "The legacies of eruption: Matching traces of ancient volcanism with chronicles of cold and famine," *The Sciences* 31, no. 1 (1991), 30–35.

These results all indicate that volcanically induced failure of the Nile summer floods might have triggered rebellions, curbed foreign wars, and potentially led to the demise of the Ptolemaic Kingdom. The end of the kingdom is typically attributed to the suicide of its last ruler, Cleopatra, in 30 BCE following her defeat by Rome. But it also followed a decade with two major volcanic eruptions (46 and 44 BCE), which led to repeated Nile failure and a societal cataclysm that included famine, plague, corruption, and migration. Environmental triggers of societal demise, such as volcanic eruptions, however, are best discussed with nuance against a backdrop of demographic, socioeconomic, and political changes. The web of interactions between humans and the environment leading to societal collapse is intricately woven. This was starkly illustrated a few centuries after the decline of the Ptolemaic dynasty, when the Roman Empire began its own lengthy implosion. No socioecological web is as labyrinthine as the one that led to the fall of Rome.

Eleven
Disintegration, or The Fall of Rome

I took Latin in school. Six endless years of Latin, from which I absorbed a sum total of two things: that Julius Caesar considered the Belgians to be the bravest among the Gauls (*Horum omnium fortissimi sunt Belgae*) and that he cast the die (*Alea iacta est*) when he crossed the Rubicon in 49 BCE. My teachers can attest that I was never an enthusiastic Latin student, and I had zero intention of incorporating anything about the Roman Empire in my future professional life. Yet, even though my career as a dendrochronologist seemed as far removed from Caesar as I could imagine, I should have known that in the end all roads lead to Rome.

My journey to the Eternal City started at the WSL in Switzerland, during my work extracting climatic information from archeological wood dating back to Roman times. By using more than 8,500 tree-ring samples from subfossil wood, timber from historic buildings and Roman water wells, as well as living oak and pine trees, our team developed a precipitation and a temperature reconstruction for central Europe that covered the past 2,400 years (405 BCE– 2008 CE).* When we lined up the tree-harvest dates of the archeological material, we found an active construction period, indicated by many felled trees, from ca. 300 BCE to 200 CE (see fig. 7). This time frame coincides with the Roman Climate Optimum, a period when the Roman agrarian economy thrived, the population prospered, and the empire reached its peak complexity against a backdrop of a generally benign European climate.

---

*We used oak samples from low-lying areas in Germany and northeastern France, where water availability is the primary limiting factor for tree growth, to reconstruct central European rainfall. To reconstruct summer temperature, we used pine samples from the Austrian Alps, where trees are more sensitive to temperature variability than to water availability.

How Climatic Instability Changed Europe
500BCE–2008CE

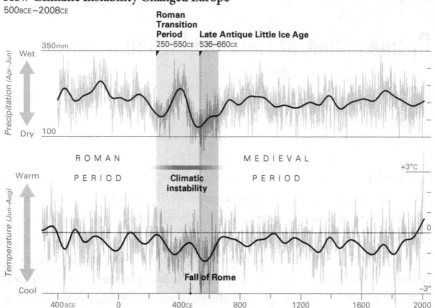

Figure 16 Around 250 CE, the Roman Empire began to experience a sequence of alternating dry and wet decades combined with consistently cool summers. This climatic instability during the Roman Transition Period continued for 300 years and thus overlapped with two events that altered Europe forever: the disintegration of the Western Roman Empire and the *Völkerwanderung*, the Migration Period, during which Germanic tribes and Huns wandered into the Roman Empire and contributed to its demise.

However, the wet, warm, and above all stable climate regime of the Roman Climate Optimum came to an end around 250 CE and was followed by an extended period of climate vagaries. Alternating dry and wet decades were accompanied by cool summers that hit rock bottom around 550 (fig. 16). These three centuries of exceptional climatic instability coincided with an important transition in the Roman Empire. With the size of the empire growing out of control, it was split into its western and eastern halves in 285, which divided its strengths and largely dissolved its coherence. Rome ruled the Western Empire, Constantinople the East. When the Germanic king Odoacer invaded Rome almost 200 years later, deposing the last western emperor, Romulus Augustus, the Western Empire fragmented and went into freefall.

Over the course of 300 years, the empire transitioned from a sociopolitically complex state that embraced a multitude of regionally diverse cultures to an amalgam of rump states, decapitated after the fall of its capital. The timeline of the failure of the Roman state is fairly well established and accepted, thanks to the Romans' love of writing. The circumstances contributing to its disintegration, however, have long been debated among historians and archeologists. There is no consensus about the relative role of internal failures, such as escalating corruption and civil war, versus external factors, such as the barbarian invasions and pandemics. Our climate reconstruction showed that this Roman Transition Period coincided with severe European climate instability, which raises the possibility that climate also contributed to the disintegration of the Roman state.

When studying potential links between climate history and human history, one of the most important principles to keep in mind is that correlation does not mean causation. To make a case for the role of climate vagaries in the failure of the Roman state, we need to establish realistic pathways through which an unstable climate could have interacted with political drivers and societal vulnerabilities to disrupt Roman society and to create a synergism that made the implosion of the existing sociopolitical system all but unavoidable.

The first of three potential pathways is probably the most intuitive one: the decadal fluctuations in hydroclimate and the cool temperatures of the Roman Transition Period were detrimental to agricultural productivity. The Roman Empire expanded over three continents (Europe, North Africa, and Southwest Asia) and comprised a wide range of climate systems, offering the empire some resilience to run-of-the-mill weather whims. This geographical diversity provided a buffer to occasional local irregularities but was not up to the task of mitigating the large-scale climate turbulence of the Roman Transition Period. While widespread summer cooling shortened the growing season and reduced harvests in Europe, droughts diminished Rome's granary in North Africa. According to papyri data, during the Roman Climate Optimum the Nile produced a fruitful flood on average once every five years, but during the Roman Transition Period favorable Nile floods occurred less than once a decade. Such decadal climate fluctuations are disruptive for agricultural societies because they are difficult to fight with social or technological innovations; this is true even in complex societies such as those the Romans had built. More frequent, annual fluctuations can be mitigated by building up, storing, and using reserves. But there are only so many reserves one can store. Once periods of drought exceed more than 5 to 10 years, conditions for food production and society as a whole can become dire.

The role of Rome's societal fabric in enabling and aggravating the impact of climate perturbations on the agricultural engine of its economy cannot be emphasized enough. The million-resident metropolis of Rome relied on a relatively small farming community to feed a large nation of urbanites and army men. Toward the time of its fall, Rome's administration counted more than 35,000 employees, and its army numbered more than 500,000. The empire comprised more than 1,000 cities, all of which relied on surrounding farmlands for supplies and support. In lean years with low agricultural productivity, it was the rural community that suffered most from food shortages and starvation; this had a crippling effect on farmworkers and led to further productivity cutbacks. To make matters worse, late Roman society was not only top heavy but self-indulgent.* Rome's ruling class enjoyed their wine and olives. Because the best and most productive farmlands were reserved for such profitable crops, staples such as wheat and barley were pushed to more marginal lands. Agriculture in marginal environments is less productive, of course, but it also carries more risk: marginal farmland is more sensitive to climate change. The tumultuous climate of the Roman Transition Period disproportionally impacted the yield of drought-sensitive staples on these marginal lands, which plummeted, reducing the carrying capacity of the Roman agrarian economy.

A second potential connection between climate instability and the disintegration of the Roman state is through the *Völkerwanderung*, the Migration Period, from ca. 250 to 410 CE, during which Germanic tribes, such as the Saxons, Franks, and Visigoths, migrated to and infiltrated the Roman Empire. In 410 CE, they invaded the city of Rome. The barbarians migrated westward into the Roman Empire because they were fleeing the Huns, who in turn were migrating westward from Central Asia. One theory postulates that the westward migration of the nomadic Huns might have been driven by drought in their original stomping grounds that disrupted their pastoral economy. To investigate the validity of this theory, my LTRR colleague Paul Sheppard and his collaborators developed a 2,500-plus-year-long tree-ring chronology from

---

* Any resemblance to contemporary societies is (not) entirely coincidental.

drought-sensitive Qilian juniper (*Sabina przewalskii*) trees on the Tibetan Plateau. To achieve such an extraordinarily long tree-ring chronology, they sampled 800-year-old living juniper trees, as well as historical construction wood. They then extended their tree-ring chronology even further back in time by sampling wooden coffins found in subterranean tomb chambers dating back to the seventh through ninth centuries. Their Tibetan Plateau tree-ring chronology reveals a severe drought in Central Asia in the fourth century that might have led the nomadic Huns to quite literally seek greener pastures to the west and south. In doing so, the Huns invaded barbarian grounds, and the rest is history.

In early January 2011, I embarked on my own personal migration period, relocating from Switzerland to Tucson, Arizona. Right before my move, the journal *Science* had accepted our findings on climatic instability during the Roman Transition Period for publication. I was not closely tracking the publication process of our paper during my hectic transition, but my new employers at the University of Arizona and their press office were. In my first week in Tucson, my drives around town to obtain such necessities as a cell phone and a bank account were complicated by the press office's barrage of phone calls. Thanks to their professionalism and perseverance, we managed to put together a press release just in time for the January 12 publication date.

And then Gabby Giffords was shot.

On a sunny Saturday morning, five days after I moved to Arizona and four days before our paper release, the Arizona congresswoman was at a grocery store in northwest Tucson meeting with constituents when a man ran up and shot her in the head. He then turned his weapon on the crowd and shot 19 more people. Giffords survived, but 6 others did not. The attempted assassination of a US congresswoman at a public event was a stark and immediate confrontation with the gun laws of the community I had just joined. I wasn't in Switzerland anymore. That next week, the mass shooting was the headline in all state and national news outlets, and I learned that even the most exciting scientific story sometimes must be set aside for more shocking or urgent news.

Shortly after I started my new job in Tucson, David Soren, a famed classics professor and archeologist at UA, sought me out. Despite the lack of broad publicity, he had read our paper. He pointed me to a third potential avenue linking the disintegration of the Roman Empire to climatic change: epidemic diseases. Dr. Soren handed me a pamphlet he had written entitled *Malaria, Witchcraft, Infant Cemeteries and the Fall of Rome*. We engaged in a lively discussion about the potential links between our research findings, and despite my aforementioned resistance to the dulcet tones of The Scorpions, I have been striving to come up with equally catchy titles for my writing ever since.

Since the late 1980s, Soren has led the excavation of a Roman villa near Lugnano, in Umbria, that was destroyed in the third century CE. It was repurposed in the mid-fifth century as an infant cemetery, La Necropoli dei Bambini. DNA analysis has shown that the 47 infants interred in the cemetery, all under the age of 3, were victims of a malaria epidemic. The excavation further revealed telltale signs of magic practices: puppies with severed heads, a raven talon, and toad bones. These macabre objects suggest that the Romans in fifth-century Lugnano employed witchcraft to banish the evil malaria demon, even though the Roman Empire was nominally Christian by that time. More evidence of witchcraft was found when the remains of a 10-year-old child possibly infected by malaria were discovered at the site. As part of the funeral ritual, now referred to as a "vampire burial," a rock had been inserted into the child's mouth to prevent it from rising from the dead and spreading malaria to the living.

Malaria gets its name from the Italian phrase *mala aria*, which means "bad air" and traces back to the Roman belief that malaria was caused by the pungent air that rises from swamps and marshes. In Roman times, the deadly disease was common in the Mediterranean region. It was most prevalent during harvest time, in late summer and early fall, when it would confine farmers to bed, while the fields lay neglected. This debilitating effect on farmworkers' health undoubtedly affected agricultural productivity and food production. Could the climate shenanigans of the Roman Transition Period have amplified the role of malaria in the fall of Rome? Possibly. The alternating wet and dry decades of the third through sixth centuries, combined with widespread deforestation, would have created optimal conditions to produce more marshy environments and thus more breeding sites for the mosquito vectors of malaria, more infected rural farmers, and less food for urbanites and paisans alike.

European summers were generally cool throughout the Roman Transition Period, but the cold temperatures spiraled even further in the two centuries following the fall of the Western Roman Empire in 476. The Late Antique Little Ice Age (LALIA) was a strikingly cold period from 536 until around 660 that enveloped the entire Eurasian continent. The centuries of LALIA cold are captured by our 2,500-year-long temperature reconstruction from the Austrian Alps and also in an almost equally long (359 BCE–2011 CE) summer-temperature reconstruction for the Altai Mountains in western Russia, more than 4,700 miles to the east. The LALIA started with a bang in 536, a year so cold that the Irish Annals recorded "a failure of bread," and John of Ephesus, a contemporary writer in Constantinople, observed that "all the wine had the taste of reject grapes."* For a long time, the cause of the 536 cold extreme was unclear and a topic of intense scientific discussion. The few surviving documentary records for this year are generally vague and have generated potential explanations that included volcanoes, interstellar clouds, and asteroid or comet impacts. John of Ephesus, for instance, also wrote that "the sun became dark and its darkness lasted for eighteen months. Each day it shone for about four hours, and still this light was only a feeble shadow. Everyone declared that the sun would never recover its full light." The Byzantine historian Procopius recorded that "the sun gave forth its light without brightness . . . and it seemed exceedingly like the sun in eclipse, for the beams it shed were not clear."†

The enigma of the 536 abrupt cold spell was resolved when Michael Sigl and his colleagues used the 775 radiocarbon spike to line up volcanic eruptions in ice-core records with cold years in tree-ring records. When they closed the seven-year gap between the ice-core and the tree-ring records, it became blatantly obvious that 536 was the year of the first of a series of three volcanic eruptions that occurred over a short period of time. The rapid succession of two large and one minor volcanic eruption had triggered a decade-long string of cold summers that kicked off the LALIA. The Alps and Altai tree-ring re-

---

* Pseudo-Dionysius of Tel-Mahre, Chronicle, 65.

† Procopius, *History of the wars*, trans. H. B. Dewing (Cambridge, MA: Harvard University Press, 1916), 4.14.5.

cords show that summer temperatures in the 540s were 3.4 to 5.8 degrees Fahrenheit colder than average throughout Eurasia.*

The 536 event was most likely a severe northern high-latitude eruption, but its precise geographical origin is not yet known. It was followed four years later in 540 by a massive eruption of tropical origin, likely the Ilopango volcano in current-day El Salvador, even larger than the one in Tambora that triggered the 1816 "Year Without a Summer." We know this because it was recorded in ice cores from both hemispheres (in Greenland and Antarctica) as well as in frost rings in the ancient bristlecone pines and in narrow ring widths across the globe (Ireland, Europe, Russia, Argentina). The 536 and 540 eruptions created a thick volcanic dust veil that blocked the sun, cooled Earth's surface, hindered plants in their photosynthesis, and threatened food security. The reports on this from contemporary writers are supported by stable carbon-isotope measurements in subfossil wood from northern Finland. The carbon-isotope measurements, which reflect changes in solar radiation, show extreme reductions in solar irradiance in the rings for 536 and 540. By comparison, the 547 eruption was smaller, but still substantial, and the three-volcano cluster kicked the LALIA cooling into gear.

The LALIA was jolted into a cold phase by the volcanic triumvirate, but it was exacerbated and sustained over the following century by a solar minimum and by negative North Atlantic Oscillation (NAO) conditions. A 3,000-year-long stalagmite record from the Uamh an Tartair cave in Scotland (the cave we used in our NAO reconstruction) shows a switch from positive to negative NAO conditions around 550. The North Atlantic pressure difference that whirled Atlantic Ocean warmth toward Europe before the LALIA stalled and left the continent open to cold air moving in from Siberia in the east.

The harsh cold at the start of the LALIA hit a Roman Empire already weakened by the preceding centuries of civil war, *Völkerwanderung*, and the damage inflicted by climate instability on Rome's agrarian economy and social order. By the time of the 536 eruption, the solidity of the Roman Empire had been broken, and its western half had crumpled under the combined pressure of harvest failures, epidemics, and barbarian invasions. The Eastern Roman Empire was also faced with the double whammy of the LALIA and a disas-

---

*This temperature difference is calculated relative to the 1961–90 reference period.

trous plague in the sixth century but nevertheless survived for a much longer period of time, until its defeat by the Ottoman Empire in 1453.

Shortly after the 536 and 540 eruptions, the bubonic plague reached the eastern shores of the Roman Empire from the Asian highlands and spread westward throughout the empire as a pandemic of unprecedented caliber. The plague first reached the shores of Egypt in 541, just one year after the second volcanic eruption, on rat- and flea-infested grain ships. Egypt's granaries fed the booming rat population, which swiftly diffused throughout the Roman world. In 542, plague-infected rats were brought to Constantinople, which had become the empire's new capital after Rome's demise, on grain ships arriving from Egypt. From Constantinople, the plague spread to port cities throughout the Mediterranean, and by 544 it had reached the western edge of the empire, in the British Isles. The combination of rodent-infested infrastructure and global trade created a perfect storm for the rapid-onset yet long-lasting pandemic we now refer to as Justinian's plague.

The plague festered throughout the empire for two centuries. Its final outbreak occurred in the 740s, after which the pandemic ended as abruptly as it had started. The fact that the pandemic began mere years after the first volcanic spasms of the LALIA and ended as the LALIA gave way to medieval warmth raises questions about potential links. Bubonic-plague epidemics result from a complex synergy of biological and environmental factors. In his 2017 book *The Fate of Rome: Climate, Disease, and the End of an Empire*, Kyle Harper outlines how the expansive infrastructure network of ships, cities, and granaries in the late Roman Empire created a landscape conducive to pandemic plague. Harper describes the plague epidemic as a biological domino event that involved at least six different species: the plague bacterium (*Yersinia pestis*) itself, the fleas that carried it, the gerbils, marmots, and black rats that were bitten and infected by the fleas, and finally humans, who contracted the illness through flea bites or contact with rats. Changes in temperature and rainfall can affect the habitat, behavior, and physiology of each of the organisms involved in the plague cycle. Climatic changes can therefore amplify or repress the epidemic, depending on where and when in the chain reaction they occur. The potential link between climatic change—such as the cooling of the LALIA—and the plague is thus complex and nonlinear. The most likely scenario involves the mid-sixth-century switch to negative NAO conditions, which increased rainfall in semi-arid Asia, *Yersinia's* homeland, and boosted regional gerbil and marmot populations. This boost in the sylvatic (wild animal) host population increased the interface with other hosts, such as black rats, which in turn traveled on board the many trade ships bound for the Roman Empire.

The concatenation of the LALIA cold and Justinian's plague took an excessively heavy toll on the late Roman population, which was already frail coming out of the Roman Transition Period. Plague mortality rates throughout the empire are estimated at 50 to 60 percent. Such immense loss of life sent Roman society into a deep downward spiral. The abrupt depopulation of half of the farmworkers and soldiers resulted in harvests rotting in the field, food scarcity, and a military crisis. It speaks for the resilience of the Eastern Roman Empire that it survived not only this disruptive transformation of its socioeconomic system but also the rise of the Islamic civilization toward the end of the LALIA. The Eastern Roman Empire recovered from two centuries of plague pandemics and mid-seventh-century Muslim conquests, reemerging by the tenth century as the flourishing Byzantine Empire, which dominated the eastern Mediterranean political and cultural spheres for centuries to come.

Twelve
It's the End of the World as We Know It

The Khorgo lava field, the volcanic field at the base of the Khorgo volcano, which lies at the heart of the Terkhiin Tsagaan Nuur National Park in Mongolia, is filled with solidified lava bubbles that are referred to as "basalt yurts." In addition to being a popular local tourist attraction, the lava field is a textbook example of where to find old, climate-sensitive trees. This region in central Mongolia is arid, receiving less than 10 inches of precipitation per year. The black basalt field creates microsite conditions where sparse trees grow at a very slow pace. Like the dolomite outcrops in the American Great Basin that harbor the ancient bristlecone pines, the lava flow on the slopes of the Khorgo volcano has very little or no topsoil and thus few wood-decaying microorganisms and bacteria. Dead wood, standing and fallen, can remain on the Khorgo landscape for centuries, even millennia.

I interviewed Amy Hessl, a dendrochronologist from West Virginia University who has worked extensively on the Khorgo lava field, over skype. It was February and from my sunny porch in Tucson, I could see the snow falling through the window of Amy's house as she told me that during the 2010 Mongolia field campaign her colleagues "thought she was crazy."

One of Amy's colleagues on the campaign, Neil Pederson of Harvard Forest, had sampled living Siberian larch (*Larix sibirica*) trees at Khorgo on an earlier trip. The oldest tree cored on that trip had been about 750 years old. Amy had suggested revisiting the Khorgo lava field in search of even older trees and dead wood. Her colleagues were reluctant, because this would involve a harrowing twenty-hour round trip on bad roads from where they were stationed in Ulaanbaatar, Mongolia's capital. Nevertheless, she persisted.

The beginning of the trip to Khorgo did not inspire confidence. On the way from Ulaanbaatar to Khorgo, Neil had risked eating a questionable wild mushroom stew in a local restaurant. By the time the team arrived in Khorgo, he was sick as a dog and had to bow out of the first day of fieldwork. So on that first day, Amy visited the Khorgo site with two Mongolian students, who both refused to bring water. Used to drinking hot tea with yak milk, they were convinced that they could work a full day on a hot basalt lava field without water. But the sun quickly caught up with them, and they became severely dehydrated. With only Amy's water for the three of them, they had no choice but to return to the campsite without having sampled a single tree. Fortunately, things looked up the next day. Now equipped with sufficient provisions and reunited with Neil, the team took a different trail and before long stumbled upon a lava wasteland where sparse, stunted, clearly old Siberian pines (*Pinus sibirica*) were growing and snags and logs littered the landscape. The team spent five more days collecting more than 100 samples at the Khorgo site before returning to the US.

Amy and Neil arrived home right before the start of fall semester, when their heavy teaching loads forced them to neglect the hard-earned Khorgo samples. "Neil and I did not crossdate any of that material for eight months," Amy told me. "Then Neil texts me one night at 8:00 or so with just a number." That number was 657, for 657 CE, the year to which he had just dated a piece of Khorgo wood. From there on out, they moved quickly. Before long, they had assembled a 1,112-year-long tree-ring chronology. It was the longest record of year-to-year drought variability on the Asian steppe.

With the Khorgo drought reconstruction in hand, the first period they focused on was the early thirteenth century, when Genghis Khan rose to power. Focusing on this period in Mongolian history was an intuitive and obvious choice for the research team. "If you were working in Mongolia, and you had a time machine, this is the period you would zoom into, to figure out the climatic backdrop of one of history's great events," Amy told me. Genghis Khan was proclaimed the universal leader of the Mongols in 1206.* He spent the next two decades leading one successful military campaign after another until his death in 1227. Under his leadership, the Mongols conquered a vast area covering most of Central Asia and China.

The tree-ring time machine that Amy and Neil had developed told them a clear story: Genghis Khan built and expanded his empire during the wettest

---

*"Universal leader" is the literal translation of *Genghis*.

decades of the past 1,000 years. The period from 1211 to 1225, the apex of Genghis's conquests, stands out in the tree-ring record as 15 consecutive wide rings. This section represents a 15-year-long period of above-average rainfall, or a pluvial, that knows no parallel over the 1,112-year record. The most intuitive link between the wet and warm climate of the early thirteenth century and Genghis Khan's success is that grasslands flourished under these conditions and provided all the fodder Genghis could want for his growing cavalry.

Cavalry was at the center of Mongolian military tactics. Mongolian horses were small, at ca. 50 inches being closer in height to ponies than to many other horse breeds, but they were the Mongols' "war machines," indispensable for the mounted cavalry archers. Mongolian archers were some of the best horsemen in the world, and history has it that they would slide down the side of their horses to shield their bodies from enemy arrows. Hanging sideways and still at full gallop, an archer would hold his bow under and parallel to his horse's chin to return fire. Genghis Khan himself is alleged to have claimed that conquering the world was easy from the back of a horse.

The Mongolian war machine thrived under the early thirteenth-century pluvial, which improved grassland productivity in the arid Mongolian steppe. The benign climate also increased the carrying capacity of the land and thus favored the concentration of resources and the centralization of power. In 1220, 10 years into the pluvial, Genghis Khan established a small military outpost in Karakorum, at the edge of the Orkhon Valley. The development of the Karakorum outpost into a political and military center required a concentration of people, armies, and horses that would have been unthinkable under less fortunate climate conditions in a strictly pastoralist society, where surplus resources are nonexistent. The favorable climate of the early thirteenth century, however, allowed Genghis to consolidate the political and military might of the Mongol Empire and feed its rapid expansion.

Yet, the unification of a massive empire made up of hundreds of individual tribes required more than just increased land productivity and energy availability. It also required a charismatic leader and the right socioeconomic and political circumstances for such a champion to rise. This is where the Khorgo drought reconstruction sheds additional light on Mongol history. An exceptional drought in Genghis's early lifetime, from the 1180s until the turn of the thirteenth century, preceded the 1211–25 pluvial. This severe drought coincided with political turmoil in Mongolia, which included relentless internal warfare and the disintegration of existing hierarchies. It was against this backdrop of societal destabilization that Genghis Khan rose to power and first unified the Mongol Empire.

From the Mongol episodes of a drought in the 1180s and a pluvial in the second decade of the thirteenth century, it would be easy to conclude that unfavorable climates are connected to societal destabilization and that favorable climates help the rise of great empires. But even though climatic instability has been linked to many a societal transformation, as dendrochronologists have demonstrated time and again, climatic changes most often constitute just one component in a web of interlinked contributing factors, as we saw in the case of the Roman Empire. By themselves, climatic changes cannot explain the downfall of civilizations. Whether or not climate change will lead to the disintegration of existing societal structures is determined by many factors, most important of which are the vulnerability, resilience, and adaptive capacity of a society itself. Compounding external factors, such as epidemics and competing societies, can also play a role. If and how a society responds to the imminent threat of catastrophe hinges on its cultural values and how they are reflected in its socioeconomic structure, such as the top-heavy structure of the Roman Empire, and its political leadership. We are perhaps experiencing an even more compelling example right now: for the first time in history, our scientific methods are sufficiently advanced to forecast in detail the threat of global, man-made climate change, but our inability (or unwillingness) to mitigate or act upon this knowledge is largely owing to political decision making, or the lack thereof.

Recent research advances in dendrochronology have helped to highlight the pivotal impact of a society's response to environmental threats on its chances of withstanding and overcoming them. The precision of tree-ring-based climate reconstructions that cover well-documented periods in human history has empowered us to devote the close attention to past human-environment interactions that they need and deserve. By extending their Khorgo tree-ring chronology even further back in time, for instance, Amy and her collaborators succeeded in providing a climate timeline for an earlier episode in Mongolian history, in the eighth and ninth centuries, approximately 450 years before

Genghis Khan's rise, when the Uyghur Empire thrived. To do this, they organized three additional field campaigns to the Mongolian Plateau to collect more material, including from a second lava flow, Uurgat, close to the capital of the Uyghur Empire. The resulting Mongolia tree-ring chronology combines even more samples from living trees, snags, and logs and now extends back 2,700 years to 688 BCE. The extended Mongolia tree-ring chronology provides a reliable drought reconstruction that illuminates the rise and fall of the Uyghur Empire from 744 to 840 CE.

The Uyghurs were a nomadic steppe people who replaced the Turks as the rulers of Inner Asia in the 740s. Their economy was primarily pastoral, but it was diversified and sophisticated, allowing them to develop a strong communication and trade network between China, Central Asia, and the Mediterranean. Upon coming to power, the Uyghur leadership established a symbiotic relationship with the Tang leadership in China, which involved trading Uyghur military power and horse production for Chinese silk, the most lucrative asset of the day. In this way, the Uyghur Empire situated itself as a major player on the Silk Road and soon stretched from the Caspian Sea in the west to Manchuria in the east. The first half of the Uyghur period, from its start in 744 until 782, was characterized by mild and moderately wet conditions that benefited Uyghur horse populations, and therefore its silk trade with China and its sophisticated pastoral economy. However, the favorable climate disintegrated in 783, ushered in by the start of a 68-year drought. From the get-go, this drought disturbed the fabric of the Uyghur Empire, but it took almost its full seven-decade length to bring the empire to its knees in 840.

At its onset, the drought brought a period of political instability, which included a war with Tibet (789–92), followed by an interruption of the horses-for-silk trade with China from 795 to 805.* The drought then intensified, reaching its nadir between 805 and 815, the very driest decade of the drought era. Despite this extreme drought, the Uyghurs managed to restore their trade relationship with China. We know from contemporary Chinese texts that the Uyghur horse trade with China culminated in the late 820s, when the Uyghurs

---

* It is unclear whether this interruption was caused by lower horse production, the army's increased need for horses, the increased difficulty of long-distance horse transport under drought conditions, or a combination of all of these.

traded 5,750 horses in 829 and 10,000 horses in 830 for a total of 230,000 silk cloths. The sustained, even increased Uyghur commerce attests to the empire's resilience in the face of persistent climatic stress. By diversifying its economy and shifting the focus from one of four pillars—pastoralism, agriculture, trade, and military service—to another, the Uyghur Empire managed to mitigate the most adverse societal effects of environmental stress for a full generation.

Eventually, however, the relentless drought proved too much even for the sophisticated economy of the Uyghurs. As the drought devastated grass and horse productivity, the inflow of Chinese silk itself ran dry after 830, ruining the Uyghur economy. The economic collapse was swiftly followed by political conflicts. In 839–40 a *dzud*, a catastrophically cold winter accompanied by heavy snowfall and high livestock mortality, dealt the final blow to the Uyghur Empire. Dzuds are not recorded in the tree-ring record—trees grow in summer, after all—but historical records reveal the extensive loss of livestock, the sweeping epidemics, and the pervasive famine that ruled that winter. The Kirghiz, a people in southern Siberia subject to the Uyghur Empire, saw this crisis as an opportunity to rebel. They invaded the empire, destroyed the capital, killed the Uyghur emperor,* and ended an almost century-long reign. The immediate causes of the decline of the Uyghur Empire—the economic crisis, the dzud, the Kirghiz—have long been understood, but the Mongolia tree-ring chronology now shows that they arrived in the wake of more than half a century of drought. The relentless drought contributed not only to the cessation of trade with China but also to the economic and political crisis following the catastrophic 839–40 dzud.

▪ Roughly 2,500 miles south of where the Uyghur crisis was taking place, another power emerged simultaneously in continental Southeast Asia: the Khmer Empire. The remains of its capital, Angkor, in current-day Cambodia, are among the most inspirational and important archeological sites in the world and have been recognized as a UNESCO World Heritage Site. A thousand years ago, Angkor was a sprawling urban complex and an extensive hydraulic city. Angkor's urban center was connected with a vast network of suburbs and agricultural land through a sophisticated water-management system that included canals, dikes, and reservoirs. The Angkor hydraulic network covered almost 400 square miles

---

*Or *khagan* in the Turkic and Mongolian languages.

designed to distribute the rainfall delivered by the summer monsoon. In most years, the summer monsoon brings moist air and rain from the Indian Ocean to Southeast Asia, producing peak rainfall in July and August. The hydraulic city of Angkor was well adapted to the monsoon rainfall regime as long as it was reliable but proved very vulnerable to abrupt shifts.

To study the history of the Southeast Asian summer monsoon and its impact on the Khmer Empire, Brendan Buckley, of the Lamont-Doherty Tree Ring Lab at Columbia University, and colleagues sampled the rare cypress *Fokienia hodginsii* (or *po mu*) in Vietnam. Like many other tropical tree species, fokienia shows irregular stem growth, and many of its rings are missing over parts of the stem or are interspersed with false rings. This makes the species difficult to crossdate, but it is very valuable because of its longevity and because it is a good recorder of drought. By taking as many as seven samples per tree, rather than the standard two, and by discarding samples that did not crossdate, Brendan managed to develop a 750-plus-year-long fokienia tree-ring chronology that captured summer monsoon variability during the Khmer Empire.

Brendan's monsoon drought reconstruction (1250–2008) shows that the East Asian summer monsoon became very fickle in the decades leading up to the fall of Angkor in the fifteenth century (fig. 17). The abnormally weak mon-

Megadroughts in Southeast Asia
1250–2008CE

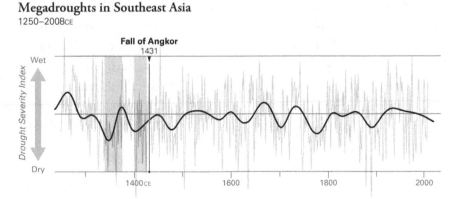

Figure 17 The East Asian summer monsoon became very fickle in the decades leading up to the fall of Angkor. A ca. 35-year drought in the mid-fourteenth century was occasionally interrupted by intense monsoon floods and followed by a shorter but at times more severe drought in the early fifteenth century. Angkor's hydraulic infrastructure was not fit to handle such abrupt switches from drought to flood and vice versa.

soon in the mid- to late fourteenth century resulted in an approximately 35-year drought (1340–75), which was occasionally interrupted by abrupt and intense monsoon floods. As we saw in the case of the Roman Empire, such abrupt switches from drought to flood and vice versa are difficult for societies to deal with. As it turns out, Angkor's famed sophisticated water network failed under the abrupt and extreme monsoon floods. Because of its size and complexity, the Angkor hydraulic infrastructure was unwieldy and hard to alter, with many potential points of failure. It was not fit to handle the fourteenth-century monsoon shenanigans.

Evidence for the damage to Angkor's water-management system exists, not through dendrochronology but in the surprising form of 650-year-old leaves. The delicate material was recovered from sediment that had been deposited in one of Angkor's main canals, and radiocarbon dating places it in the late fourteenth century. The preponderance of fallen leaves suggests that the canal was filled at the time of Angkor's demise with flood sediment that had eroded from surrounding areas. Such sediment filling would have stopped the canal from bringing irrigation water from the hinterland to Angkor's urban core. The abrupt monsoon floods appear to have damaged Angkor's hydraulic infrastructure right in the middle of a 35-year-long drought period, when the surrounding agricultural land most needed irrigation and flood control.

Angkor's fourteenth-century monsoon debacle did not take place in a vacuum. Again, socioeconomic and geopolitical turmoil, which included an escalating war with the neighboring Ayudhaya Kingdom, accompanied the climatic challenges. When the fourteenth-century monsoon drought was followed by a shorter but at times more severe drought in the early fifteenth century, Angkor proved fatally weakened by the continued climatic, socioeconomic, and political death spiral and eventually fell in 1431. Only its temple complex, Angkor Wat, has survived to this day as a Buddhist monastery, the largest religious monument in the world.

▧ Dendrochronology has also shed light on the devolution and drivers of the Mayan *Terminal Classic Period* in the eight through tenth centuries, which is captured by the demise of their *Long Count calendar*. The Maya were one of the most advanced societies worldwide during the Classic Period (ca. 250–750 CE). Their extensive cities, beautified by opulent art and architecture, were home to millions of people, who recorded their daily life and adventures in a hieroglyphic script. They used the Long Count calendar to identify a date by counting the number of days that had passed since the creation of the world, which, according to Mayan creation myths, had occurred more than three thousand years earlier.* Not many of these Mayan written records survived the sixteenth-century Spanish Conquest, but a multitude of inscriptions on buildings and monuments are still visible and decipherable today. This is particularly true for the calendar dates the Mayans sculpted on new buildings, the earliest of which dates back to 197 CE. During the Terminal Classic Period (ca. 750–950 CE), however, one by one Mayan cities stopped erecting dated monuments. The last preserved Mayan Long Count date is 909, found in the city of Tonina, in Chiapas.

The termination of 700 years of Long Count dates appears abrupt, but the disintegration of Mayan society that it represents was a circuitous process that stretched out in time; it wasn't a sudden or complete demise. Between 90 and 99 percent of the Mayan population is estimated to have been lost during the Terminal Classic Period. However, hundreds of thousands of Mayans were still left to fight the Spanish conquistadors when they arrived in the sixteenth century, incurring an additional massive population loss, though even this second dark night of the Mayan soul did not ultimately eviscerate the society, and today the Mayan population in Meso-America has rebounded to 6 or 7 million people. Despite the perseverance of the Mayan people over the subsequent centuries, it is hard to dispute the disastrous effects of massive population loss and the decline of Mayan institutions, along with the disappearance of its kings and its Long Count calendar during the Terminal Classic Period.

The precise cause of the Mayan disintegration is debated, but the hypothesis that climate change played a role is more than a century old. It was originally posited by Ellsworth Huntington, a Yale geography professor and contemporary of A. E. Douglass's. Huntington flirted heavily with the ideas of *climatic determinism*, even though this eighteenth-century approach had largely fallen from favor by the early twentieth century because of its association with scientific racism, colonialism, and eugenics. Nevertheless, Huntington postulated that a Meso-American pluvial provoked the Terminal Classic Mayan decline,

---

* The creation date was 31 August 3114 BCE, according to the Gregorian calendar.

and he supported his faulty hypothesis using sequoia tree-ring data from California. Huntington hypothesized that climate variability in the Yucatán was inversely related to that in California: when it was dry in the Yucatán, it was wet in California and vice versa. According to Huntington, a tenth-century drought recorded in Californian tree-ring data thus corresponded to a pluvial in the Yucatán.

Huntington's idea to use tree-ring data to investigate the Mayan Terminal Classic Period was good, but his execution was poor. Rather than tree rings from remote California and a hypothesized inverse climatic teleconnection, the rings in millennium-old Meso-American trees would prove enlightening. Few Meso-American tree-ring chronologies extend back far enough to study pre-Columbian history. Because much of the Meso-American landscape has long been denuded and developed, tree-ring chronologies typically do not extend beyond 300 to 400 years. But in the steep Barranca de Amealco gorge, about 50 miles north of Mexico City, Jose Villanueva Diaz, of the Instituto Nacional de Investigaciones Forestales, Agrícolas y Pecuarias (INIFAP) in Durango, Mexico, and Dave Stahle, of the University of Arkansas, found a grove of Montezuma bald cypress (*Taxodium mucronatum*) that had escaped centuries of human deforestation. Ironically enough, the Montezuma bald cypress is related to the North American sequoia, so in a way, Huntington had been on the right track.

The Montezuma bald cypress is the national tree of Mexico and the only Meso-American tree species that lives to be 1,000 years old. One Montezuma bald cypress specimen in Oaxaca is, at 37.5 feet in diameter, the largest tree in the world by circumference. The trees in Barranca de Amealco are not quite as stout, but at 14 feet in diameter, they still provide a 1,238-year-long tree-ring chronology that reflects past droughts in central Mexico. The Amealco tree-ring chronology (771–2008) documents at least four major droughts, the first of which lasted 25 years (897–922) during the Terminal Classic Period. The chronology narrows the timing of the Terminal Classic drought down to the year and shows that drought, rather than Huntington's pluvial, bedeviled the region in the tenth century. It also shows that the drought extended over a broad geographical range from the Yucatán into the highlands of central Mexico. The tree rings further recorded two later droughts that corresponded to the timing of the demise of the Toltec state (1149–67) and to the sixteenth-century Spanish conquest of the Aztec state (1514–39), respectively.

Like the earlier Terminal Classic drought, this sixteenth-century drought coincides with a massive depopulation in Mexico that has been scrutinized in both the contemporary scientific world and the popular media. The Aztec population is estimated to have been decimated by 80–90 percent over the course of a century after Europeans arrived. European and African diseases imported by the conquistadors, such as smallpox and measles, contributed to this sixteenth-century demographic catastrophe, but its main culprit was an endemic disease that the Aztecs named *cocolitzli*, meaning "pestilence." Cocolitzli was a hemorrhagic fever that was unknown to both European and Aztec physicians. It was likely caused by a viral infection, not unlike the Ebola and Marburg viruses, that raged through the Aztec population in a series of epidemics, while leaving the Spaniards all but unaffected. A first cocolitzli epidemic started in 1545, 24 years after the Spanish conquest of the Aztec Empire, lasted four years, and killed 800,000 people in the Valley of Mexico alone. An even larger cocolitzli epidemic started in 1576, resulting in the deaths of 45 percent of the remaining Aztec population.

The two sixteenth-century epidemics occurred during an almost century-long drought that lasted from 1540 to 1625 and extended from Central Mexico through North America to the boreal forest of Canada. A close look at the Amealco tree-ring chronology reveals that the 1545 and 1576 cocolitzli epidemics started during brief wet periods that interrupted the persistent decades of drought. The parallels between the sixteenth-century and Terminal Classic droughts and their respective devastating population losses are striking. They suggest that hemorrhagic-fever epidemics might have been involved not only in the sixteenth-century Meso-American depopulation but also in the Mayan demise of the Terminal Classic Period. Such alternating dry and wet episodes have been linked to pandemics on other occasions as well, such as the Hantavirus outbreaks in the American Southwest in the 1990s. We have also seen them at work on decadal scales, when they provided a climatic link between malaria outbreaks and the fall of Rome.

Working with Villanueva Diaz and Stahle, Matt Therrell, of the University of Alabama, and Rodolfo Acuna-Soto, of the Universidad Nacional Autónoma de México, have since expanded upon the Amealco tree-ring chronology, so that the Mexican tree-ring network now includes more than 30 Douglas fir and Montezuma bald cypress chronologies. This network has been used to lend credibility to some pre-Hispanic Aztec beliefs. The Aztecs were a superstitious people; they held folklore high and believed in omens and curses. Arguably the most infamous of these curses was the "Curse of One Rabbit," which predicted famine and devastation in each year of "One Rabbit," the first year of the 52-year Aztec calendar cycle. To test the validity of the Curse of One Rabbit, the researchers looked at tree-ring-reconstructed drought before, during, and after One Rabbit years. Lo and behold, they found that no fewer than 10 of the 13 pre-Hispanic One Rabbit years on record (from 882 to 1558) followed severe droughts in the preceding year. The widely reported Famine of One Rabbit in 1454, for instance, was preceded by two years of below-average tree-ring values, indicating drought and potential crop failure. The Aztecs might have been on to something in relating their One Rabbit years to famine and misfortune, but their curse ended with the Aztec Era. None of the 8 One Rabbit years recorded after 1558 CE and following the Spanish Conquest were preceded by low tree-ring values or drought. Cocolitzli appears to have not only decimated the Aztec population but also wiped out its Rabbits.

Thirteen
Once upon a Time in the West

Two years after I moved to Tucson, my home at the University of Arizona's Laboratory of Tree-Ring Research moved to a brand-new, specially designed building on campus after 75 years in its "temporary" location under the bleachers of the UA football stadium. Over those 75 years, the LTRR collection of tree-ring samples had grown exponentially, and the space under the football stadium was bursting at its seams. Five years later, at the time of this writing, we are still in the process of moving the collection, which comprises more than 700,000 samples.

Of those, 400,000 belong to the southwestern dendroarcheology collection. With the earliest sample dating back to 171 CE and the most recent cookie cut in 1972, this collection tells us the stories of 1,800 years of southwestern US history. The tree-ring samples share with us intricate details about the lifestyle of the Ancestral Puebloans,* the climatic conditions under which they lived, and how the two are linked. Like most other preindustrial civilizations, the Puebloans used wood profusely and consistently as a building material, to make artifacts, to cook, and to heat their homes. In the cold-steppe environment of the Four Corners region of Colorado, New Mexico, Arizona, and Utah, much of that wood has been preserved through time.

Prior to Douglass's discovery of tree-ring dating in 1929, however, the age of the Ancestral Puebloan ruins in the Four Corners was largely unknown. Archeologists debated the age of the grand pueblos and cliff dwellings that nowadays are preserved in places such as Mesa Verde National Park in south-

---

*The Ancestral Puebloan archeological culture has long been referred to as the Anasazi culture. The term *Anasazi*, which means "ancestors of our enemies," was introduced by the Diné (Navajo) people, who moved into the Four Corners region around 1400 CE. Out of respect for contemporary Puebloans, I here use the term *Ancestral Puebloans*.

ern Colorado and Chaco Culture National Historical Park in northern New Mexico. For instance, in 1922 the head of the archeological mission at Chaco Canyon estimated that its Pueblo Bonito complex had been occupied "say 800 or 1,200 years ago."[*] When Douglass dated the most recent tree ring in Chaco Canyon exactly to 1132 CE, he unleashed a chronological beast that altered the world of southwestern archeology and anthropology forever.

The majority of Ancestral Puebloan archeological wood samples that are sent to the LTRR for dating are charcoal remnants of buildings and structures that burned or of cooking and heating fires in hearths. Charcoal is more likely to be well preserved than regular wood because it consists almost entirely of carbon. It therefore lacks the cellulose and sugars that insects and microbes like, and that lead to decay when wood is buried by soil. The primary structural aspects of wood are preserved when it is charred, and charcoal fragments can therefore show clear tree rings with earlywood and latewood and even specific anatomical characteristics, such as resin ducts, that can be used to identify the tree species. A charcoal fragment needs to show enough of these clear rings, however, for it to be crossdated successfully. If the fragment is small or derived from a fast-growing tree with wide rings, as is often the case for charcoal found in hearths, it may only have twenty or even fewer rings. This is not enough to reliably match the charcoal's tree-ring pattern to a unique spot in the reference tree-ring chronology and thus to date it. As a result of such short tree-ring sequences, only about 40 percent of all the archeological samples from the American Southwest that are archived in the LTRR have been dated successfully.

Nevertheless, it's fair to say that the number of southwestern charcoal fragments that have been tree-ring dated at the LTRR has grown steadily over the years. According to my colleague Ron Towner, the LTRR archive contains more than 100,000 dated charcoal fragments. They are complemented by a lower number of uncharred archeological wood samples, mostly derived from wood beams. Wood beams that have remained above ground for their entire existence have, like charcoal, mostly escaped wood-decaying microorganisms and can be preserved for more than 1,000 years. Combined, the charcoal and wood samples elucidate the chronology and environment of the Ancestral Puebloan culture.

[*] Neil M. Judd, "The Pueblo Bonito expedition of the National Geographic Society," *National Geographic Magazine* 41, no. 3 (1922), 323.

Chaco Canyon, in Chaco Culture National Historical Park, was one of the bulwarks of the Ancestral Puebloan culture. The canyon is about 15 miles long and up to a mile wide, and both of its walls are lined with large, multi-storied pueblos. From the mid-ninth to the mid-twelfth century, Chaco Canyon was a major center of culture, politics, and commerce. Its role as a regional and cultural hub was supported by public and ceremonial architecture that included twelve great houses and many more kivas, underground round rooms used for religious rituals, political meetings, and community gatherings. The largest and most studied great house at Chaco Canyon is Pueblo Bonita, which covered almost two acres, was up to four stories high, and comprised more than 650 rooms. The construction of the urban ceremonial center of Chaco Canyon is estimated to have required more than 200,000 timbers. The Chacoans used logs to roof their great houses and kivas, as bonding beams inside masonry walls, and as support pillars. Smaller pieces of wood were used for roofing and flooring and as lintels and sills.

Nowadays, a visit to Chaco Canyon is a trek. It is located in a remote corner of northwestern New Mexico, and on the three-hour drive there from Albuquerque or Santa Fe, you can't help but wonder why anyone would choose to settle in this desolate landscape. There is no hotel, but if you want to stay overnight to see the sun rise or set over the ruin-clad canyon walls, there is a campground. You'll have to bring your own firewood because the Chaco landscape is largely treeless. There are a few sparse stands of pinyon pine and juniper and the occasional stunted ponderosa pine or Douglas fir, but wood is hard to come by. This is what the Chaco landscape has looked like for the longest time. Large pine, spruce, and fir trees have largely been absent since the end of the Pleistocene, 12,000 years ago.*

Up until the early eleventh century, population density in Chaco Canyon was low and the Chacoans relied primarily on local wood resources—pinyon, juniper, and cottonwood—for their construction needs. But as the settlement expanded, more and more great houses and large kivas were built, and the gnarly little trees that surrounded the canyon could not provide the long, straight roof timbers needed for such majestic structures. The few ponderosa pines and Douglas firs in Chaco Canyon that could fulfill this function were

---

*We know this from the analysis of tree pollen in ancient, radiocarbon-dated packrat middens.

cut early in Chaco's development, so that by the mid-eleventh century the Chacoans were forced to import wood from surrounding mountain ranges. The remote origin of the Chacoan timber is corroborated by the scarcity of stone axes (the main instrument for tree cutting) found over more than a century of archeological excavations in the canyon itself. The surrounding mountain ranges, by contrast, are covered with axes and choppers.

More unequivocal evidence for the source of Chaco timber is provided by dendroprovenancing. Most of the southwestern archeological wood in the LTRR archive is crossdated visually, without measuring the ring widths. Ron Towner and his colleagues know the centuries-long sequence of wide and narrow rings (the "Morse code") of the southwestern tree-ring chronology by heart. Often they can date a charcoal fragment just by looking at its ring-width pattern. Jeff Dean, southwestern dendroarcheologist *extraordinaire* and an LTRR colleague, tells me that his go-to tree-ring signature for dating southwestern samples is from the 1250s: the rings for 1251, 1254, and 1258 are typically narrow, and the ring for 1259 is wide. If Jeff sees this signature in an undated sample, it gives him a place to start; he then can work from there to make sure that the rest of the sequence matches as well. This method has allowed Jeff, Ron, and their colleagues to date more than 100,000 southwestern samples over the past 90 years. However, to determine the origin of Chaco's timber, Chris Guiterman, a PhD student at the LTRR, measured each tree-ring width on a subset of 170 beams from Chaco great houses, a much more painstaking process than visual crossdating. He then compared the ring-width patterns in the beams with those found in the tree-ring chronologies from eight potential harvesting areas in the mountain ranges surrounding Chaco Canyon. He found that 70 percent of the measured Chaco timbers originated from the Chuska Mountains, 50 miles west of Chaco Canyon, and the Zuni Mountains, 50 miles to the south.

The slopes of the Chuska and Zuni Mountains are forested with large, mixed conifer forests that produce the long and straight beams needed for the construction of great houses and kivas. To harvest these perfect spruce, fir, and pine beams, Chacoans walked the 50 miles from the canyon and back. They had no wheels or horses for transportation, and their main tool for cutting down the large trees was the aforementioned hand-held stone axe. It doesn't take much imagination to see that tree harvesting took an enormous amount of time and energy. It took about 100 person-hours to haul a single beam from the Chuska Mountains back to Chaco Canyon, and more than 2,000 person-trips to build the largest great houses, such as Pueblo Bonito. Yet this did not stop the Chacoans from importing tens of thousands of beams to construct their cultural metropole.

Chris further detected a striking change in the preferred wood source region. Prior to ca. 1020, the majority of timber was imported from the Zuni Mountains. Less than 50 years later the Chuska Mountains had eclipsed the Zunis as the primary harvesting area. This temporal shift from the Zunis to the Chuskas coincides with the mid-eleventh-century onset of Chaco florescence. During the accompanying surge in construction in the second part of the eleventh century, seven new great houses—half of the total in the canyon—were built and most existing great houses were expanded.

Despite the superhuman efforts the Chacoans invested in pursuing building material and constructing their monumental great houses, they occupied the buildings for a remarkably short time. Only a hundred years after its mid-eleventh-century peak, Chaco Canyon was all but deserted. Mere decades after hauling tens of thousands of beams from the Zuni and the Chuska Mountains, the Chacoans packed up their bags and left. Having built the largest structures in pre-Hispanic North America, costing them more than 1 million person-hours, they ended up occupying them for only a few generations.

Similar stories can be found all over the Ancestral Puebloan Southwest. The cliff dwelling of Betatakin, for instance, a Kayenta construction built in the late thirteenth century, was occupied for less than 40 years. When I asked Ron Towner about this remarkably short return on investment, he replied that perhaps the largest revelation dendrochronology has offered to southwestern archeology, next to absolute dating, has been the evidence that the occupation of most of the Ancestral Puebloan structures was fleeting. As Ron puts it, "It wasn't until the advent of dendrochronology that people realized that just because these structures have been standing there for 800 years, does not mean that they were occupied for 800 years."

The corollary of such short occupation times is that the Ancestral Puebloan culture was characterized by mobility. If you ask twenty-first-century descendants of the Ancestral Puebloans about the depopulation of Chaco Canyon, they aren't surprised. It was time to leave, that's all. Time to dismantle the Chacoan experiment, with its complex organization and society. Chacoans dispersed throughout the region, and many returned to a more mobile lifestyle, with smaller, shorter-lived structures and less formalized, centralized groups.

The growth and decline of Chaco Canyon exemplifies a demographic cycle that is an inherent part of Ancestral Puebloan history as we understand it. The cycle starts with a long period of exploration, during which Puebloan populations dispersed and explored potential new settlement locations and new forms of organization. This exploration phase is gradually replaced by an exploitation phase, when some of the exploration trials were successful and allowed the explorers to settle down and invest energy in agriculture and in the building of great houses and kivas. This exploitation phase is then followed by a relatively rapid disintegration of the achieved aggregation, after which a new cycle starts with slow exploration. The whole cycle takes one to two centuries, and then starts again.

This demographic cycle of societal expansion and contraction shows up in the tree-ring record. Kyle Bocinsky, a computational anthropologist affiliated with Crow Canyon Archeological Center, in southwestern Colorado, and his collaborators compiled almost 30,000 tree-harvest dates covering the period from 500 to 1400 CE from more than 1,000 archeological sites in the Four Corners region. When they lined up how many trees were harvested per year (a similar effort to what we did for Roman timbers in central Europe), they were astonished to find that the tree-harvest dates clustered in four distinct peaks that were separated by four quieter intervals (fig. 18). The peaks were the results of building frenzies during exploitation phases, whereas dips corresponded to exploration phases, with less construction. They found that each exploitation peak lasted about a century, was preceded by an exploration phase with a slowly increasing number of tree-harvest dates, and ended in an abrupt decline in tree harvesting. The four exploitation peaks Bocinsky found in the Ancestral Puebloan record dated to 600–700, 790–890, 1035–1145 (the Chaco peak), and 1200–1285. This ultimate tree-harvest spike, the Mesa Verde peak, corresponds to the flourishing of the Mesa Verde and Kayenta cultures in the thirteenth century.

Unlike previous construction peaks, however, the 1285 Mesa Verde peak was not followed by an exploration phase with a slowly increasing pulse of

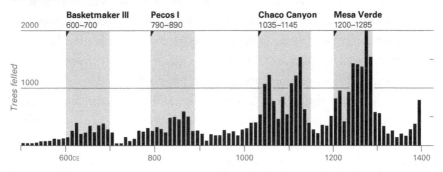

Figure 18 By comparing the harvest dates of nearly 30,000 trees in the Four Corners region from the years 500–1400 CE, we can see four distinct peaks in building activity. Each peak lasted about a century and ended in an abrupt decline in tree-harvest dates.

tree-harvest dates. Rather, the number of tree-harvest dates dwindled after 1285 and did not recover. The 1285 deterioration event appears to have been the most drastic one. Previous cultures, including the Chacoan culture, had concluded their demographic cycles by leaving their settlements to move around the region. The Mesa Verde culture, however, left the Four Corners region entirely in 1285 and did not return. The mobility and settlement cycle that defined Ancestral Puebloan societies was sustainable only as long as population densities were low and there were unoccupied areas to explore and occupy. But by the late thirteenth century, people filled most of the landscape in the region, and when it came time for the Mesa Verde residents to leave, the only options left were to reoccupy previously abandoned land or to move out and south. Most Mesa Verdians moved southward to the Mogollon Rim and the San Juan Basin in present-day New Mexico. The Mesa Verde Ancestral Puebloans did not vanish after 1285; rather, their people and culture were integrated into other southwestern societies, whose descendants still live in Hopi and Zuni pueblos in New Mexico today, and visit the dwellings of their ancestors for religious ceremonies.

But there was more to the story of the Mesa Verde exodus than increasing population densities alone. With overpopulation came overexploitation of natural resources, and this proved to be fatally detrimental in the fragile environment of the American Southwest. Both water and wood are scarce resources in the arid land of the Four Corners region, and they are dangerously easy to overexploit, as proved the case for the ponderosa pines and Douglas firs in Chaco Canyon, leaving the landscape largely treeless for centuries to come. The desolate landscape you drive through on your road trip to Chaco Culture National Historical Park is a mute witness to the overexploitation that occurred a thousand years ago.

Like the Mayan disintegration 2,000 miles to the southeast and 300 years earlier, the Four Corners overexploitation did not happen in isolation. Rather, it was superimposed on severe and persistent droughts that hit Chaco Canyon in the 1130s and the Kayenta and Mesa Verde cultures in the 1280s. Such *megadroughts*, which lasted 20, 30, or even 50 years, would have made dryland and irrigation agriculture impossible and food storage, which can cover a few years but not decades, futile. These megadroughts were unlike anything we have witnessed in recent years, even during the Dust Bowl of the 1930s or the Sahel drought of the 1970s and 1980s. Societies with lower population densities and living in less denuded landscapes might have found a way to deal with such challenging conditions, but the Kayenta and Mesa Verde societies, living on densely populated and heavily developed lands, had no choice but to leave.

How do we know about these megadroughts? From tree rings, of course. The late thirteenth-century Mesa Verde megadrought has been part of the history of dendrochronology since its very beginning. Back in 1935, Douglass wrote: "The great drouth from 1276 to 1299 was the most severe of all those represented in this 1200-year record and undoubtedly was connected with extensive disturbances in the welfare of the Pueblo people."[*] As we saw in chapter 1, Douglass was the first to provide absolute, dendrochronological dates for archeological sites across the Four Corners region. It took Douglass almost 15 years of sampling and crossdating countless tree-ring samples to bridge the gap between his living-tree chronology and his floating (relatively but not absolutely dated) archeological chronology. In 1929, he bridged the gap between the two chronologies with sample HH-39. The gap centered on 1286 (see fig. 1), right in the middle of the "great drouth," as Douglass named it. There

---

[*] A. E. Douglass, *Dating Pueblo Bonito and other ruins of the Southwest*, Pueblo Bonito Series, no. 1 (Washington, DC: National Geographic Society, 1935), 49.

were two main reasons why it took Douglass so long to bridge this gap. For one, many of the Puebloan sites in the Four Corners region, including Mesa Verde and Kayenta, were abandoned around that time, and the amount of archeological wood available for sampling and analysis dropped significantly. But the co-occurring late thirteenth-century Mesa Verde megadrought also contributed to the gap. The few samples that were available for the late thirteenth century were replete with missing or microscopic rings, which greatly hindered crossdating. In Douglass's December 1929 article in *National Geographic*, "The Secret of the Southwest Solved by Talkative Tree Rings," he described the late thirteenth-century sequence of rings in sample HH-39, including the 1250s tree-ring signature Jeff Dean was referring to, as follows: "Following its rings inward toward the core, we saw the record of the great drought. Here were the very small rings that told of the hardships the tree had endured in 1299 and 1295. As we studied the rings further toward the center, 1288, 1286, 1283, and 1280 each told the same story we had found in other beams of lean years and hard living. Also, there were the years 1278, 1276, and 1275, the ring for each corroborating the diary entries other logs had given us. . . . Here was its account of 1258, a hard year, and of 1254, an even harder one. Presently it told of 1251 and 1247, years when all the trees were singing 'How dry I am.'"

Initially Douglass's hypothesis that the abandonment of Mesa Verde and other sites was related to the great drought received criticism. Southwestern archeologists were quick to note that Douglass's ponderosa pines were primarily sensitive to winter moisture, whereas maize, the Ancestral Puebloans' main crop, was grown in summer. Such summer agriculture was made possible by the North American summer monsoon, which contributes up to half of the total annual rainfall in much of the American Southwest. To address this issue of winter versus summer drought, Dave Stahle, who was also involved in the discovery of the Montezuma bald cypress grove in Barranca de Amealco and of the ancient bald cypresses in North Carolina, came up with the idea of measuring earlywood and latewood widths separately in southwestern tree-ring series. He found that in some southwestern tree species there is a very distinct boundary between their earlywood, which is influenced by winter precipitation, and their latewood, which responds to the summer rains brought by the North American monsoon.

To find trees old enough to tell the story of the Great Drought in the late thirteenth century, Dave went to the El Malpais National Monument in New Mexico, where he found living trees and remnant wood in a lava-flow environment similar to the one that Amy Hessl and colleagues sampled in Mongolia. The resulting El Malpais chronology covers more than 2,000 years, and Dave and his team spent more than a month measuring earlywood and latewood widths in the El Malpais samples. Their efforts have resulted in two separate precipitation reconstructions, one for winter precipitation based on earlywood widths and one for summer-monsoon precipitation based on latewood widths. Dave's work shows that the thirteenth-century Great Drought was indeed primarily a winter event, whereas the 1950s southwestern drought, for instance, was shorter in overall duration but lasted throughout both winter and summer.

▨ Since Douglass's bridging of the gap, hundreds of tree-ring chronologies have been developed for the American Southwest, where the science is almost a century old, the trees are long-lived and drought-sensitive, and the scientists are hardy. The southwestern tree-ring network includes, among others, the 8,000-plus-year-long bristlecone pine chronologies and the blue oak chronologies from the Central Valley in California—the best recorders of drought in the world. Over time, precipitation-sensitive tree-ring chronologies have also been developed across the rest of North America and most of the extratropics. Ed Cook, of the Lamont-Doherty Tree Ring Lab at Columbia University, has used the North American section of this tree-ring network to develop the North American Drought Atlas.* This network of 2,000-year-long reconstructions has been two decades in the making, and the latest version provides details of past drought conditions for every point in North America on a 0.5-degree latitude/longitude grid. Ed and his team have since developed similar drought atlases for Europe (the Old World Drought Atlas), Monsoon Asia, Mexico, and Australia. A seasonal drought atlas, in which summer and winter drought are reconstructed separately using Stahle's earlywood-latewood technique, is in the works.

The single greatest event in the North American Drought Atlas is the twelfth-century Chaco megadrought, which is followed closely by the thirteenth-century

---

*See http://drought.memphis.edu/NADA/.

Mesa Verde, or Great, Drought (fig. 19), but the atlas also allows us to study the spatial extent of the megadroughts. Thanks to the atlas, we now know that the Chaco and Mesa Verde events were not restricted to the Southwest but were widespread throughout the entire American West. The Chaco and Mesa Verde megadroughts happened during the Medieval Climate Anomaly, when Northern Hemisphere temperatures were about 1.3 degrees Fahrenheit warmer than during the subsequent Little Ice Age, as we know from the Hockey Stick and the Spaghetti Plate studies. In Europe, medieval warmth helped the Viking expansion and British viticulture. In the American West, warm temperatures help explain the prevalence of medieval megadroughts. Drought, after all, is a function not only of how much water enters the earth system as rainfall or snowfall but also of how much water leaves the system through evaporation and transpiration. In that sense, the earth system is not different from the human body. Think of how much more you sweat, and how much easier it is to become dehydrated, when you go for a hike on a hot day than it is when you hike on a cool day. In the same way, even if the amount of rainfall stayed the same from the Medieval Climate Anomaly into the Little Ice Age, the warmer medieval temperatures would have led to longer and more severe droughts.

While medieval times were warm, their temperatures have been exceeded in recent decades under anthropogenic climate change. We are direct witnesses

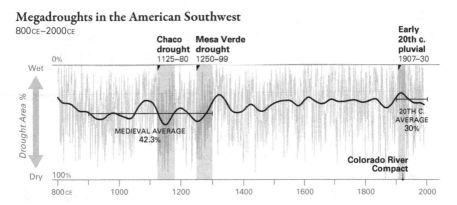

Megadroughts in the American Southwest

Figure 19 The single greatest megadrought in the North American Drought Atlas is the Chaco megadrought around 1150. The wettest period in the past millennium occurred in the early twentieth century and coincided with the signing of the 1922 Colorado River Compact, an interstate water-rights agreement based on less than 30 years of data.

to the impacts of this recent warming in the American West. California endured a five-year drought in 2012–16 that culminated in the 500-year record Sierra Nevada snowpack low my colleagues and I uncovered. Up until 2018, the Southwest experienced two decades of drought that started in 1999 and has left "bathtub rings," marking former higher water levels, in Lake Mead and other water reservoirs. In June 1999 Arizona governor Jane Dee Hull declared a state drought emergency that is still in effect two decades later. But however severe these recent conditions seem, they pale in comparison with the medieval megadroughts. Not only did the medieval droughts last longer—up to 50 or more years—but they were more severe and widespread than the worst droughts of the twentieth and twenty-first centuries. If such megadroughts were to recur now, they would pose immense challenges to existing western water-management systems. High streamflows, which are critical for refilling the reservoirs, would be lacking for decades. If the entire American West experienced conditions akin to the medieval megadroughts, areas such as southern California, which relies on water supplies from the Colorado River as well as from Sierra Nevada snowfall, would be in big trouble.

It is disquieting to think that the medieval megadroughts were part of the natural variability of the climate system. The twelfth-century Chaco event, for instance, occurred during a period when solar activity was at a peak and volcanic activity was at a low. This led to warm temperatures that not only directly exacerbated the drought in the American West, as described above, but also had an indirect impact by promoting ocean-atmosphere dynamical patterns that are linked to drought in the Southwest, such as the La Niña phase of the ENSO system. There is every reason to believe that such natural variations in the climate system will recur in the future, and when they do, they will be layered on top of man-made warming. In short, the climate system can create severe droughts all by itself, but our recent warming of the atmosphere, population growth, land-use changes, and the related overallocation of water resources increase the chances of such droughts being "mega."

In developing water-management plans in western North America, taking into account the information that tree rings provide about the long-term context of megadroughts is key. We have learned this the hard way from the Colorado River Compact. This interstate compact, drawn up in 1922 by the Colorado River Commission under the chairmanship of Herbert Hoover, lays down the "Law of the River" for the division of the waters of the Colorado River among seven western states and Mexico. For the purpose of the compact, the Colorado River basin was divided into the Upper and the Lower Basin. The dividing point was set at Lees Ferry, in northern Arizona, which today is the launch point for tourist boats into the Grand Canyon. The negotiators of the compact used the river gauge at Lees Ferry as a baseline to determine the amount of Colorado River water available per year for allocation to the Upper and Lower Basin states. It is not fully clear how much water the compact negotiators thought they had to work with, but it was on the magnitude of 16 to 17 million acre-feet (MAF) per year, with 1 acre-foot roughly the equivalent of the amount of water that will support four families in the Southwest for a year. What is clear is that the negotiators of the compact felt comfortable enough with the annual Colorado River flow to effectively allocate 15 MAF per year, 7.5 MAF to the Upper and Lower Basins each. The Mexican Water Treaty of 1944 further commits the US to send 1.5 MAF per year to Mexico, thus topping the legal entitlements to Colorado River water off at 16.5 MAF per year.

In retrospect, the timing of the 1922 compact was very unfortunate. The compact negotiators based their allocations on the couple of decades' data that were available at the time. These early twentieth-century streamflow measurements, however, were not representative of long-term Colorado River water availability. On the contrary, we know now from tree-ring data that the 1922 compact was written during one of the wettest periods in the past 500 years (see fig. 19). We can use tree rings to reconstruct streamflow in the Colorado River basin because both tree rings and streamflow are controlled by the same hydroclimatic factors, such as snowfall and evapotranspiration. In 1976, Chuck Stockton, of the LTRR, and Gordon Jacoby, of the Lamont-Doherty Tree Ring Lab, first used tree rings to reconstruct the Colorado River gauge record at Lees Ferry back to 1521. They found that the long-term average flow of the Colorado River was not the 16.5 MAF per year used in the compact's allocation but rather 13.5 MAF per year, a whopping 3 MAF, or 12 million families, per year lower. They further found that the longest period of high flow in the 450-year record occurred in the early twentieth century, from 1907 to 1930, right when the 1922 compact was drafted. Stockton and Jacoby's original reconstruction has since been refined and extended with more tree-ring data, and the longest Lees Ferry streamflow reconstruction now dates back to 762 CE. The four or five Lees Ferry streamflow reconstructions currently in existence do not all agree on the average flow of the Colorado River, and estimates range from a low of 13 MAF to a high of 14.7 MAF. But even the best-case 14.7 MAF average annual flow is well below the Colorado River allocation, the equivalent of annual water supplies for more than 7 million families.

The Lees Ferry streamflow reconstruction demonstrates how tree-ring data provide a much-needed long-term context for the recent western North American droughts. Tree rings have shown that the actual worst-case scenario for western drought is far worse than the worst-case conditions that twentieth- and twenty-first-century water-management strategies are based on. For instance, in the twentieth century the longest stretch of time that the Colorado River went without a high flow was five years. During the twelfth-century Chaco megadrought, the longest stretch was not five years but sixty. The twenty-first-century southwestern drought, which has lasted two decades so far, is just a teenager compared with the medieval megadroughts. Imagine the bathtub rings in Lake Mead if it were to hit a full-blown midlife crisis! If we want to avoid abandonment of the American Southwest and a repetition of the ultimate outcomes of the Ancestral Puebloan experiment, our water-management plans, such as the Colorado River Drought Contingency Plan, need to be rooted in the long-term context of megadrought provided by tree rings and other paleoclimate data. This will help us to manage our western water in a sustainable way so that future populations, cities, ecosystems, and dendrochronologists can continue to thrive in this environment.

Fourteen
Will the Wind Ever Remember?

Earth's climate is a complex system. We are experiencing that firsthand through man-made climate change. Physical laws dictate that rising greenhouse-gas emissions lead to rising temperatures—literally global warming. In practice, it looks more like global weirding, which includes heat waves but also 20-year droughts, wildfires, category 5 hurricanes, the polar vortex, and "snow-mageddons." Such climatic diversity and complexity is not well represented by global climate averages, such as the Hockey Stick, alone. Fortunately, the global tree-ring network that we have been compiling over the last century or so helps us to better put the current crazy climate in a long-term context. It allows us to map spatial climate patterns and to pick and choose, to mix and match tree-ring chronologies in order to look at dynamical, rather than average, climate patterns. For our North Atlantic Oscillation (NAO) reconstruction, for instance, we compared the Moroccan Atlas cedar tree-ring record to a Scottish stalagmite proxy. When we make such connections between climate proxies, the cogs of the NAO might emerge from our tree-ring archive, or other dynamical aspects of the climate system, such as the El Niño Southern Oscillation (ENSO). We can also kick this kind of research up a notch and look at aspects of the climate system that do not occur at the earth's surface but in the upper layers of the atmosphere, such as jet-stream dynamics.

The *jet stream* describes the fast westerly* winds that flow five to nine miles above the earth's surface, the altitude at which airplanes fly. It is the reason why eastward transatlantic flights from North America to Europe are about an hour shorter than flights in the opposite direction. On eastward flights, pilots can fly with the jet stream and pick up speed from it. On westward flights, they have to

---

*Because of Earth's rotation.

navigate above it in order to avoid strong jet-stream headwinds. I lit upon the idea of using tree rings to reconstruct variations in the jet stream when I saw that the narrowest ring in a Bulgarian tree-ring chronology I had developed was the ring of 1976, the coldest year on record in the Balkans.

Our tree-ring chronology was based on samples from Pirin National Park, a UNESCO World Heritage Site in southeastern Bulgaria. We visited the Pirin mountain region, with its vibrant folklore and the steep, dark mountains typical of the Balkans, in 2008, after receiving a tip from Momchil Panayotov, a colleague at the University of Forestry in Sofia, who had spotted old pines on a camping trip. We assembled an international team of nine dendrochronologists from Bulgaria, Switzerland, Germany, and Belgium to collect tree-ring samples from the old Bosnian pines—the same species as Adonis, which grows about 300 miles away at the same elevation. By now you'll know the dendro-fieldwork drill: after a couple of hours' hike in the morning from the campground where we had set up camp to the treeline, we spent the day coring trees and then hiked back down in the late afternoon, reaching camp before nightfall. At the time, the Pirin National Park was a protected area, so we were not allowed to carry in a chainsaw to sample deadwood. A decade later this has changed. In December 2017, the Bulgarian government legalized commercial logging and allowed the development of a ski resort within the park's boundaries. This ecosystem threat sparked a wave of protest in early 2018 involving thousands of local and international environmentalists, but no resolution to protect the area has been agreed upon to date.

One of the park's main tourist attractions is Baikushev's pine, which is generally assumed to be Bulgaria's oldest tree. At about 1,300 years old, the Bosnian pine, which is named for its discoverer, forest ranger Kostadin Baikushev, is thought to have been around at the establishment of the first Bulgarian empire, in 681 CE. It is a stately tree, 85 feet high and more than 25 feet in circumference. We were not allowed to core Baikushev's pine. Core a national treasure? Not likely! Honestly, though, I'd be surprised if the tree were 1,300 years old. It's more likely to be a heritage tree with obvious cultural importance but not necessarily old age. Not only was the storied tree in question positioned 1,000 feet below the treeline where the oldest trees are usually found, but it does not have the stunted appearance of the really old pines we found higher up, which turned out to be up to 800 years old. The Pirin pines were thus slightly younger than Adonis and his companions in Greece but had still attained a very respectable age. Back at the lab, we developed an 850-plus-year-long tree-ring chronology (1143–2009 CE) from our Pirin sample collection.

When we measured maximum latewood density in our samples to reconstruct summer temperature, we found that the 1976 ring had very light latewood and that the summer of 1976 had been one of the coldest in 850 years in the Balkans. This struck me as odd because the summer of 1976 had been one of the warmest on record in northwestern Europe. Until the worldwide heat wave of 2018, the summer of 1976 had been the heat wave of reference where I grew up in Belgium, the heat wave with which all other heat waves were compared. When we then compared our Balkans summer-temperature reconstruction with a summer-temperature reconstruction for the British Isles, representing northwestern Europe, we saw that this disparity in the summer of 1976 had not been an exception. On the contrary, most cold spells in the Balkans over the past 300 years co-occurred with hot summers in the British Isles, and vice versa; when it was colder than normal in the Balkans, it typically was hotter than normal in the British Isles. The summer of 1976 appeared to be representative of a summer-temperature dipole between northwestern and southeastern Europe, and our tree-ring data showed that this dipole had consistently been at work for almost 300 years.

When I visited Belgium during the summer following the 2012 publication of our findings, the weather was absolutely miserable. The unseasonably nasty cold and endless rain featured prominently in the newspaper, and it was while I was reading *De Standaard* over breakfast at my parents' house that I saw The Map. I was mildly puzzled to see that the regional weather map published that day very closely echoed the Balkans–British Isles temperature dipole map we had published just a few months earlier; it showed that while we were shivering in Belgium, the Balkans were melting in a heat wave.

The accompanying newspaper article explained that this dipole pattern was a result of the jet stream's extreme southward position (fig. 20). In an average summer, the polar jet is positioned at about 52 degrees north over the eastern North Atlantic Ocean before it moves further east to just north of Scotland and Scandinavia. The polar jet can be thought of as the border between cool, Arctic air, which stays north of the jet, and warm, subtropical air to the south. In summers when the North Atlantic jet (the section of the polar

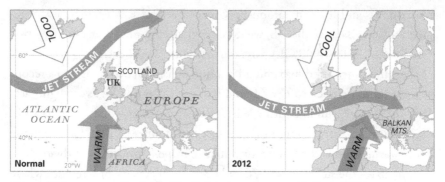

Figure 20 In summer, the North Atlantic jet stream moves at latitude 52°N on average before it flows eastward north of Scotland. It forms the border between cool Artic air to the north and subtropical air to the south, which warms European summers. However, when the North Atlantic jet stream moves further south than normal—as it did in 2012— cold Artic air descends into northern Europe. At the same time, subtropical air concentrates and creates heat waves over the Balkans.

jet stream over the eastern North Atlantic Ocean) moves further south than normal, as was the case in 2012, Arctic air and cold temperatures reach further south than normal, into the British Isles and Belgium. At the same time, warm air coming from the subtropics is concentrated over the Balkans, leading to heat waves there. The opposite pattern occurs in summers when the North Atlantic jet moves further north than normal, creating heat waves in the British Isles and relatively cool summers in the Balkans.

In some years a similar southward excursion of the polar jet stream occurs over eastern North America. This phenomenon, which brings cold Arctic air over the eastern half of the US, is referred to in the media as the *polar vortex.* Climatologically speaking, there is always a polar vortex. In the Northern Hemisphere, it is the large area of low pressure and cold air surrounding the North Pole* that lies north of the circling polar jet stream. But sometimes the jet stream surges much further south than normal, allowing the ice-cold air of the polar vortex to escape south of its normal position. Such southward excursions of the jet stream are not unusual per se. The jet stream does not circle the earth in a perfectly straight line but rather meanders around the globe like a snake. Sometimes it is strong and comes close to moving in a straight line: it moves fast, its wandering is confined, and it constrains the polar vortex to a more or less concentric region around the

---

* A similar polar vortex also surrounds the South Pole.

pole. At other times, it undulates in big northward and southward waves, reaching far northern and far southern positions. When the jet meanders widely in such a north-south pattern, it allows warm air from the tropics to move further north than is normal in some regions, while cool Arctic air (the polar vortex) moves further south than is normal in other regions. Such big curves slow the jet stream down, meaning that it stays in each northward or southward position longer, setting the stage for extreme weather. To offer a European example, when the jet stream is located over the British Isles for a couple of days, it brings rain, which is nothing special. But when it stays in its same position for weeks at a time, the relentless rain that it brings causes flooding, as happened in the summer of 2012. On the other hand, when the jet stream is in a more northern position for a couple of days in summer, all of my friends in Brussels head to the urban beach at Bruxelles les Bains. However, if it stays in that position for longer, they start complaining about a heat wave, as happened in the summer of 1976.

It was after reading in the newspaper about the curves and movements of the jet stream that I realized that the North Atlantic jet was the culprit behind the summer-temperature dipole over Europe and thus the tree-ring dipole we had discovered and that we might be able to link the two. Can we use tree-ring data from these two poles to reconstruct the jet stream back in time, I wondered? Can we use tree rings to reconstruct wind patterns that occur miles above the earth's surface? The idea was so compelling that I wrote a grant proposal to the National Science Foundation to investigate it. Convincing the NSF that your research is relevant, feasible, and urgent enough to warrant funding in a tight 15-page document is a daunting task. Running preliminary analyses and showing figures—in my case the dipole map—as proof of concept is the first step. Then you need to put together a budget showing how much your research will cost and find supportive collaborators. This all takes months, and I spent the better part of the 2012–13 academic year thinking, reading, and writing about my jet-stream project. At a friend's New Year's Eve dinner party, we went around the table, each of us predicting the biggest trend for the new year. I must have been haranguing everyone about the jet stream all night, because when it came to my turn to make a prediction, before I could utter a word all my friends shouted in unison, "THE JET STREAM!"

And in a sense I was right: the jet stream made big, slow waves in 2013, resulting in a barrage of extreme weather events across the Northern Hemisphere midlatitudes. In the British Isles, snowy, cold winter weather occurred as unseasonably late as mid-April. The storm Christoffer brought extreme flooding to central Europe in late spring. Heavy rains and floods also impacted Russia and China that summer. In July, a heat wave hit northwestern Europe, with temperatures into the 90s. In December, fierce winter storms brought heavy rainfall and floods to the British Isles. That same winter, North America experienced its own temperature dipole: while California was in the depths of its four-year drought, eastern North America got hammered by the polar vortex. It was so cold in January 2014 that the famed Niagara Falls froze over and snowfall reached as far south as Birmingham, Alabama, giving rise to the popular use of coinages such as "snowmageddon" and "snowpocalypse."

The recent rise in the number of such midlatitude weather extremes—droughts, floods, cold waves, heat waves—suggests a change in jet-stream behavior. And that is exactly what we have been seeing: the polar jet stream in the Northern Hemisphere has become wavier and slower than before, resulting in more frequent extreme jet-stream positions and more frequent extreme weather events. The increased number of jet-stream extremes in recent decades is co-occurring with drastic man-made changes in the global climate system, raising the question of whether the two are linked: are rising greenhouse-gas emissions and global temperatures causing the recent jet-stream waviness and the midlatitude weather extremes it creates? To answer this question, we need a record of jet-stream variability that extends back prior to man-made climate change, prior to the twentieth century. That is where our tree rings come in.

By combining tree-ring-based temperature reconstructions from the Balkans and the British Isles, we were able to reconstruct the European temperature seesaw for each summer back to 1725 and thus the northward and southward movements of the North Atlantic jet (fig. 21). Our Balkans–British Isles tree-ring combination captured northward as well as southward extremes of the North Atlantic jet over the past 290 years. When we looked at British summer heat waves recorded by thermometers in central England since 1659, we found that they consistently occurred when our reconstructed North Atlantic jet was further north than normal, keeping the Arctic air north of the British Isles. In contrast, cold spells in central England occurred in summers when the North Atlantic jet was further south than normal, when the polar vortex dipped southward and brought Arctic air as far south as England. The jet's southernmost po-

Varying Latitudes
1920–2010CE

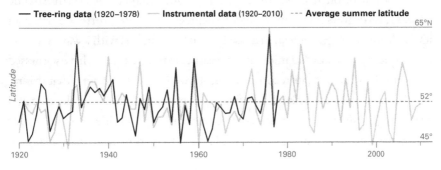

Jet-Stream Variance
1740–1997CE

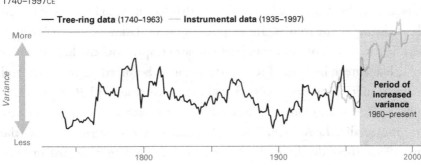

Figure 21 Using tree rings to reconstruct temperatures in Scotland and the Balkans, we were able to reconstruct the latitude of the North Atlantic jet stream back in time. On average, it has stayed at about 52°N in summer (*top*). However, since the 1960s there have been more extremes in its position (*bottom*). These northward and southward extremes cancel each other out, so the average position stays the same, but the increasing variance reflects more extreme deviations.

sition occurred in the summer of 1782, when it dipped as far south as 42 degrees north, 10 degrees (or almost 700 miles) further south than average. We know from historical documents that the summer of 1782 was so cold in Scotland that the grain harvest failed and the country was devastated by a famine.

Our reconstruction further demonstrated a trend in the frequency of jet-stream extremes (see fig. 21). The *variance* in the position of the North Atlantic jet, a metric for the number of northward and southward deviations from its average position, has been on the rise since the 1960s, showing that the summer North Atlantic jet has been reaching northward and southward extremes more often than before. This is important because it is those extreme northward and southward positions that create extreme weather events, such as the heat waves and floods mentioned above. In August 1976, for instance, the North Atlantic jet was at 65 degrees north, more than 13 degrees (900 miles) further north than its 52-degree average. Such an increase in variance is in line with a wavier jet and the frequent jet stream and midlatitude weather extremes that we have witnessed in recent decades. But our reconstruction showed for the first time that this recent increase in North Atlantic jet variance is unprecedented over the past 290 years, suggesting that the recent jet-stream extremes and waviness are not part of natural climate variability but are instead linked to man-made climate change.

The success of our North Atlantic jet-stream reconstruction emboldened us to tackle another feature of Earth's climate that has changed in recent decades: the widening of the tropics. The tropical "heart of darkness," the lush region between the Tropic of Cancer and the Tropic of Capricorn,* cinches the earth's waist like a green belt. This green tropical core is bordered on the north and south by subtropical dry zones, which are located at around 30 degrees latitude† and are home to most of the world's deserts, such as the Sahara, the deserts of Australia, the Atacama, and the Sonoran Desert near Tucson. Since the late 1970s, these dry borderlands have been expanding poleward in both hemispheres, enlarging regions parched for water.

There is a reason why the tropics are wet and the subtropics are dry: the *Hadley circulation*, an atmospheric circulation that moves warm air from the equator toward the poles. Warm, moist air rises up at the equator, where solar heating is strongest, and when it reaches about 10 miles above the surface, it starts dissipating northward and southward. As the warm air moves poleward, it cools down and rains out, thus watering the green lushness of the core tropics. By the time it has reached about 30 degrees north and south, the moving air has cooled down and dried out so much that it is no longer buoyant and starts to sink. As the cool, dry air sinks, it pushes away potential clouds and storms that bring moisture to those latitudes, leaving a desert-like landscape in its wake. Over the past four decades, the latitudes at which the tropical air sinks have moved poleward and the tropical belt has widened. This bulging of the tropical waistline can have profound impacts on the hydroclimate of the

---

* Located at 23.5 degrees north and 23.5 degrees south, respectively.
† Also called the "horse latitudes."

subtropical regions that lie just beyond them. As the dry edge of the tropical belt creeps poleward, many subtropical and semi-arid regions that used to be just beyond its reach are now smack in the middle of it. The result: drought. Southern Australia, for instance, has in recent decades been invaded by drought coming from the north. Cities just south of latitude 30 degrees south, such as Melbourne, Perth, and Adelaide, have been affected the most. In the Northern Hemisphere, cities such as Tucson (32.2°N) and San Diego (32.7°N) are in peril of losing much-needed rainfall if the tropical edge moves even as little as one additional degree further north.

Like the increase in jet-stream variance, the expansion of the tropics in recent decades coincides with anthropogenic changes to the earth's atmosphere. And as it does for the jet stream, this raises the question of whether the two are linked: does the enhanced greenhouse-gas effect cause widening of the tropics? Climate model simulations that mimic anthropogenic climate change with increased greenhouse-gas emissions tell us that yes, this is the case. But interestingly enough, the tropics expand nowhere near as fast in the climate models as they do in the real world, where they have been expanding at about half a degree (ca. 35 miles) per decade. The discrepancy between the modeled and the real world suggests that that there is more to tropical expansion than greenhouse-gas emissions alone. What more, exactly, is less clear. In the Southern Hemisphere, the ozone hole over the Antarctic may play a role. In the Northern Hemisphere, pollution from soot may be involved,[*] but the relative role of various anthropogenic drivers, as well as natural climate variability, is not fully understood. As with the jet stream, a preindustrial record of tropical-edge movements that extends back prior to man-made emissions of greenhouse gases, soot, and chlorofluorocarbons (CFCs) that caused the ozone hole, and that grasps only natural variability in tropical-edge movements could help us out. Once again, tree rings can come to the rescue.

The Argentine dendrochronologist Ricardo Villalba, at the Technological Scientific Center CONICET in Mendoza, was the first to come up with the idea of tracking past movements of the tropical edge with tree rings. Ricardo has been studying tree rings and past climate in the South American Andes

---

[*] Soot typically results from the burning of fossil fuels and biomass, including in wood stoves and forest fires. Its concentration in the atmosphere has increased substantially since 1970, particularly in the Northern Hemisphere, where there is more landmass and more burning.

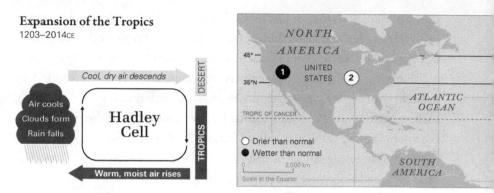

Expansion of the Tropics
1203–2014CE

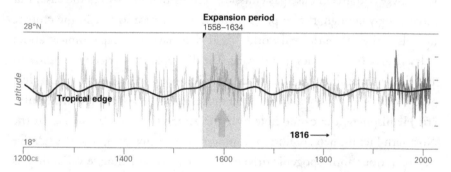

Figure 22A Regions located between ca. 35 degrees and 45 degrees north have been climatically influenced by movements of the tropical-belt boundary. When the tropical edge reaches further north than normal, some areas experience drought, while others have increased precipitation. The opposite happens when the tropical edge is located further south than normal. We have gathered five regional tree-ring chronologies in that latitudinal band: from the western US (1); the central US (2); Turkey (3); northern Pakistan (4); and the Tibetan Plateau (5).

Figure 22B By combining the five tree-ring chronologies, we can reconstruct movements of the Northern Hemisphere tropical edge back more than 800 years to 1203. We find the strongest contraction of the tropics in 1816, the "year without summer" following the 1815 volcanic eruption of Tambora. We also find a 60-year period of tropical expansion in the late sixteenth and early seventeenth centuries that was accompanied by drought and societal upheaval across the Northern Hemisphere.

since the early 1980s. At the Ameridendro conference in his hometown in 2016, he presented the idea of using his South American tree-ring network to reconstruct movements of the Southern Hemisphere edge of the tropics. Sitting in the audience that day, I was inspired by Ricardo's ideas. I asked my postdoc Raquel Alfaro-Sánchez to explore whether we could do the same for the Northern Hemisphere, whether we could use a network of Northern Hemi-

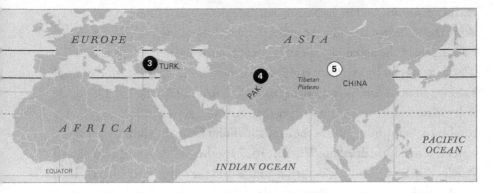

sphere tree-ring data to reconstruct past movements of the northern tropical edge. Raquel scanned the International Tree-Ring Data Bank for suitable tree-ring records from sites in the latitudinal band between ca. 35 degrees and 45 degrees north, just beyond the tropical edge. In years when the tropical edge reaches further north than normal, when the tropical waistline bulges, this latitudinal band will fall under its influence and experience drought. In years when the tropical edge is further south than normal, it will see moisture rolling in from the north and flourish. We hypothesized that drought-sensitive tree-ring data in this latitudinal band record past northward and southward tropical-edge movements. Raquel compiled tree-ring data from five regions: the western US,* the central US,† Turkey, northern Pakistan, and the Tibetan Plateau. By combining these five independent tree-ring chronologies, she developed a reconstruction of movements of the Northern Hemisphere tropical edge that extends back to 1203 (fig. 22).

When looking over the full 800 years of the reconstruction, the first prominent feature that stands out is a 60-year period of tropical expansion in the late sixteenth and early seventeenth centuries (1568–1634). Unlike the late twentieth-century tropical expansion, the late sixteenth-century expansion seems to have been a natural glitch in the chaotic climate system. To the extent of our knowledge, there was not yet an ozone hole in the late sixteenth century, and the Industrial Revolution, with its greenhouse-gas and soot emissions, hadn't started yet. But even if at this point we are unsure of its cause, the late sixteenth-

---

* These include chronologies from California, Arizona, New Mexico, Colorado, and Utah.
† These include chronologies from Arkansas, Missouri, and Kentucky.

century expansion and its accompanying drought wreaked havoc in societies all along the 35 degree latitudinal parallel (China, Turkey, the US) and serves as a warning about the potential societal impact of future poleward movements.

China was hit by a series of exceptional droughts during this period of tropical expansion. A four-year drought from 1586 to 1589 struck 350,000 square miles in eastern China, an area larger than Arizona, New Mexico, and Nevada combined. In the final years of the drought, Taihu Lake, the third largest freshwater lake in China, dried up. When in 1627 another severe drought hit Shanxi Province, west of Beijing, widespread famine and the longest peasant rebellion in Chinese history ensued. Those two droughts, however, paled in comparison with the massive drought that followed in 1638–41. Historical documents from the eastern part of Henan Province, at 33 to 35 degrees north, record horrific conditions for this period. For 1639, for instance, they report a "great drought in spring and summer, all grasses and crops withered, no reaping, the Yellow River dried up; flying locusts shut out the sunlight in spring and summer, severe famine and people ate each other in twelfth month."* The triple-drought crisis coincided with a period of economic distress, political turmoil, smallpox epidemics, and Manchu invasions, a foul brew that resulted in the wiping out of 40 percent of the population. Soon thereafter, the mighty Ming dynasty, which had ruled China for almost three centuries, collapsed.

Similar events unfolded simultaneously in the Ottoman Empire, in present-day Turkey, where a severe 1590s drought ruined harvests and led to famine and epidemic outbreaks. As we have seen many times before, sociopolitical decisions aggravated the agricultural crisis. In 1593, the Ottoman sultanate started its Long War with Austria, which demanded increased resources from the already suffering countryside to imperial Istanbul. When inflation then compounded rural shortages, desperate farmers formed rebel armies and revolted. The Celali rebellions led to the mass migration and sociopolitical destabilization that defined the early seventeenth-century Ottoman Crisis, the most severe crisis in pre–World War I Ottoman history, during which the Ottoman Empire lost almost a third of its population.

*J. Q. Fang, "Establishment of a data bank from records of climatic disasters and anomalies in ancient Chinese documents," *International Journal of Climatology* 12, no. 5 (1992), 499–515.

In North America, the late sixteenth-century tropical expansion contributed to a drought that rivaled the medieval megadroughts in intensity and far exceeded any droughts in the twentieth and twenty-first centuries. The event caused almost a dozen pueblos in the American Southwest to be permanently abandoned. It extended across the Sierra Madre Occidental, the Rocky Mountains, the Mississippi River valley, and the southeastern US. In Mexico, the megadrought coincided with the 40-year-long Chichimeca War (1550–90), the longest and most expensive conflict between European immigrants and indigenous peoples in Mexican history. It was also during the sixteenth-century expansion period that the 1576 cocolitzli epidemic contributed to massive depopulation.

Of all the drought-related calamities in this time period that are noted in American history, the most notorious is the failure of early English settlements on the East Coast. In 1587, an English settlement of about 115 colonists was established on Roanoke Island in present-day North Carolina. The colony was chartered by Queen Elizabeth I in an attempt to establish a base from which to send privateers on raids against Spain's treasure fleets. Three years later, in 1590, when an English expedition stopped at Roanoke to resupply, they found the "Lost Colony" completely deserted. There was no trace of the original 115 colonists and no sign of battle. It is assumed that the Roanoke settlers suffered through excruciating environmental circumstances that forced them to take shelter with the indigenous Croatoan tribe. It took the English 17 more years to succeed in establishing a permanent settlement in the Americas—Jamestown in Virginia—in 1607. But the Jamestown settlement also endured serious hardship and harrowing mortality rates. More than 80 percent of its original settlers perished within three years, during the Starving Time in 1609–10. The disappearance of the early English settlements is another perplexing mystery of history that tree rings have helped us solve. When Dave Stahle, at the University of Arkansas, used bald cypress tree-ring data to reconstruct drought conditions in Virginia back to 1185, he found that the Lost Colony of Roanoke had disappeared after the worst three-year drought (1587–89) in the eastern US in 800 years. Likewise, Jamestown's Starving Time occurred during the driest seven-year period (1606–12) in 770 years. The English could not have picked a worse time to come to the New World if they had tried.

For our reconstruction of movements of the tropical edge, Raquel used a network of tree-ring data spanning the Northern Hemisphere just north of this edge. In the same vein, a network of tree-ring records from the wet tropics can be used to reconstruct the ENSO system, the earth's most powerful driver of internal climate dynamics. The El Niño Southern Oscillation (ENSO) was given its name in the late 1800s by Peruvian fishermen who noticed that their Pacific Ocean fishing grounds warmed up in some years around Christmas, giving rise to the phenomenon's common name, El Niño, after the Christ child. This warming of the eastern tropical Pacific Ocean is the result of the weakening of the westward trade winds, which in normal years carry warm ocean water and moisture from South America toward the Asian tropical Pacific. We now know that every two to seven years, as these westward winds weaken, they leave a large pool of warm ocean water stuck off the Pacific coast of South America, causing tropical storms and floods. At the same time, on the other side of the Pacific, Asia and Australia receive much less moisture than normal, and experience widespread drought and wildfires. This phase of the ENSO system is referred to as El Niño. El Niño years are typically followed by La Niña years, during which the reverse happens: the westward trade winds are stronger than normal, and the warm pool, as well as the accompanying clouds and precipitation, ends up in the Western Pacific, near Asia and Australia. As a result, La Niña years are characterized by floods in Asia and Australia, but droughts in South America.

The ENSO system is responsible for the slushing back and forth of the tropical Pacific warm pool and has far-reaching hydroclimatic impacts in the regions surrounding the tropical Pacific basin. But ENSO also influences the hydroclimate of more remote regions through *teleconnection*: it links climate phenomena that happen far apart. In the Caribbean, for instance, there are typically more hurricanes in La Niña years than in El Niño years. And the only years when I can go snowboarding in the southernmost ski destination in the continental US, Mount Lemmon, near Tucson, occur when El Niño brings abundant rainfall and snow to the US Southwest. Even further afield, the flood that washed away the train tracks in Tanzania, forcing Kristof and me on our three-day-long bus journey to reach Kigoma, was caused by an El Niño event. Even though its core impacts are in the Pacific basin, ENSO has near-global teleconnection tentacles, and a better understanding of its "mood swings" is crucial for water managers around the world. This knowledge is especially important because while ENSO itself happens primarily in winter, its hydroclimatic effects often extend into the following summer. This means that more accurate ENSO forecasting could give water managers a six-month lead to prepare for upcoming droughts and floods.

By developing centuries-long records of ENSO and its teleconnections, dendrochronologists have been able to significantly improve our understanding of this climatic metronome in the tropical Pacific. Jinbao Li, of the University of Hong Kong, and his collaborators, for instance, developed a 700-year-long ENSO reconstruction (1301–2005 CE) by compiling more than 2,000 tree-ring records from both the Asian and South American sides of the tropical Pacific Ocean and from five mid-latitude regions in both hemispheres with strong ENSO teleconnections. Li's tree-ring-based ENSO reconstruction corresponds remarkably well to two coral records* from the central Pacific Ocean that reflect El Niño and La Niña phases going back to the nineteenth century. In the same way that trees form rings, corals can form yearly growth bands that record the temperature and the chemistry of the ocean water that surrounds them and can be used as proxies for ENSO variability. Only the vast reach of ENSO can make corals from the central Pacific and tree rings from Asia, South and North America, and New Zealand, all dance to the same beat.

When we compared Li's ENSO reconstruction with Raquel's reconstruction of tropical-edge movements, we found that for 700 years the Northern Hemisphere tropical belt has contracted in El Niño years and expanded in La Niña years. For instance, the late sixteenth-century tropical-expansion period, which contributed to the Lost Colony, the fall of the Ming dynasty, and the Ottoman Crisis, corresponded to a period of prevailing La Niña conditions. The ENSO and tropical-edge reconstructions have something else in common: they both show strong climatic responses to past volcanic eruptions and the aerosols they have ejected. Large tropical volcanic eruptions, such as those captured in the ice-core record, were typically followed by an El Niño year, during which the Northern Hemisphere tropical edge contracted.

The link between volcanic eruptions and tropical contractions not only helps us to understand past climate variability but also cautions us about the

---

*One from the Maiana Atoll and one from Palmyra Island.

future. The known cooling effect of volcanic eruptions has inspired *climate engineers* aiming to mitigate man-made climate change with deliberate, large-scale intervention to consider *solar radiation management* (SRM) as a potential stopgap solution for ongoing and projected global warming. Proposed SRM projects intend to artificially mimic the cooling effect of volcanic eruptions by injecting aerosols into the stratosphere from aircrafts or balloons. Whereas SRM is relatively low-cost and easy to implement compared with other climate-engineering solutions,* such as space mirrors to deflect sunlight, it fails to address many nefarious effects of increased greenhouse-gas emissions, such as ocean acidification. Furthermore, while SRM might benefit some regions and countries, it might hurt others. As paleoclimate proxies, including tree rings, improve our understanding of the impacts of volcanic eruptions on climate, the grave risks associated with SRM are also becoming prohibitively clear. Our ENSO and tropical-edge reconstructions show that stratospheric aerosols ejected by volcanoes not only cool down the earth's surface but also interfere with important atmospheric circulation systems and can result in a reorganization of precipitation and wind patterns. The tropical contractions following past volcanic eruptions suggest that artificial aerosol injection will have deleterious effects, specifically in regions such as the Middle East and the Sahel, which already are very vulnerable to hydroclimatic changes. As time progresses, and with it enhanced greenhouse effects, climate engineering is considered more and more seriously as a temporary solution until we manage to reduce atmospheric greenhouse-gas concentrations through mitigation and carbon capture. However, as the late sixteenth-century tropical expansion has shown, the societal risks involved with changing not only the temperature but also precipitation patterns on the planet are enormous.

---

*The cost of SRM is within the reach of small countries, large corporations, and even very wealthy individuals.

Fifteen
After the Gold Rush

Shortly before sunrise, around 6:00 a.m., on November 8, 2018, a utility worker for California's Pacific Gas and Electric Company (PG&E) discovered a fire burning under one of the power lines in Butte County in northern California. That morning the wind was high, with speeds approaching 50 miles per hour, the humidity was low, and the fire rapidly grew out of control. By 8:00 a.m. the fire had reached Paradise, a foothill town with an estimated population of 26,800, which was wiped off the map within a mere four hours. Because the fire spread so quickly, many residents of Paradise were unable to evacuate. At least 86 people were killed, and more than 14,000 residences were destroyed. Over the course of the 27 days that the Camp Fire raged before it was contained, it became the deadliest and most destructive fire in California's history. Today, the population of Paradise is 2,000, less than 10 percent of what it was before the Camp Fire. Throughout the 2018 California fire season, more than 8,500 wildfires charged through the state, consuming more than 3,000 square miles of towns, forests, and shrubland, costing American taxpayers $3.5 billion, half of which went toward fire suppression.

The 2018 California fire season marks a worrying trend in the destructive power, size, and economic impacts of wildfires throughout the American West. In California, 12 of the 15 most widespread fires on record have occurred since the year 2000. Across the western US, fire seasons are lasting longer, and large wildfires (those burning more than 1,000 acres) have been on the rise since the early 1980s, with every year seeing an additional 7 large fires and 90,000 more burnt acres than the last. To clarify the confluence of drivers contributing to the western US fire trend over the past four decades, such as climate change and forest-management practices, we need to understand the West's fire history and how it is linked to climate history and human history. Once again, dendrochronology can play a pivotal role in our education.

In the dry low- and mid-elevation forests of the American West, the natural fire regime consists of low-intensity *surface fires*. We know this from the scars that such groundfires leave in the stems of mature trees. Surface fires typically stay close to the ground, not reaching the forest canopy. They burn off much of the understory—grasses, shrubs, young seedlings, and saplings—but aside from some potential scarring, they are mostly harmless for large, mature trees. In fact, older trees thrive in the aftermath of surface fires, which clear out some competition for water and nutrients and reduce the risk of development of a *fuel ladder*, understory vegetation that allows fire to climb up from the forest floor to the tree canopy, becoming destructive for even the largest trees.

Many forests of the American Southwest and California harbor fire-scarred trees, living witnesses to surface fires that once burned frequently, perhaps every 5 to 10 years. Western trees can record centuries' worth of these scars; a single tree can easily carry up to 20 of them. I found my personal record number of scars on a stump in Dog Valley, near Truckee, California. My tree-ring dating told me that the stump was the remainder of a tree cut in 1854. Over the course of its 300-year life span, the tree had recorded no less than 33 fire scars.

Fire scars are often found at the base of trees that are growing on a slope and have accumulated debris—needles, branches, even logs—on their uphill side. When a surface fire moves upslope through the forest, it will linger in these uphill spots, and the fuel it finds there can make the fire locally hot and intense. It can burn through the bark of the tree, leaving an open, triangular-shaped scar known as a *cat face*.* Unlike humans, trees don't have a mechanism to heal wounds. When trees are wounded, the best they can do is form new wood and bark on either side of the wound that slowly grows over it and eventually will close it off. However, with fires occurring every 5 to 10 years, the wound typically does not have enough time to be fully sealed before the next fire arrives. The location of the original fire scar on the tree is very vulnerable to subsequent burning and wounding because it lacks protective bark and because of the high resin content of the wound tissue. When the next fire comes along, the same spot on the stem often gets wounded again. This subsequent fire then leaves a second scar, which widens the cat face and makes it even more difficult for the

---

*It is unclear where this name comes from; the scars look nothing like cat faces.

tree to grow over and cover the wound. When a third fire hits a few years later, another scar is added, the cat face opens up even further, and the pattern continues. Whenever a tree is burned by a successive fire, a new scar is added that can be tree-ring dated. Each fire scar occurs within a specific tree ring. By cross-dating the tree-ring sequence of the sample against a regional tree-ring chronology, we can date each scar to the exact calendar year when the fire happened (fig. 23). Even better, by looking at the position of the scar within a tree ring—does the scar occur in the earlywood, in the latewood, or at the boundary between two rings?—we can get an idea of the season (spring, summer, or fall, respectively) when the fire burned.

Fire-scarred trees are mostly sampled with a chain saw. Starting with a plunge cut, a skilled sawyer can remove a wedge from a living tree that covers no more than 10–20 percent of the stem's surface and removes a part that has already been heavily damaged by fire. It is not a pretty sight, but it does not hurt the tree much. On tree stumps and downed logs, where the cat face is often the best-preserved part because of the resin in the wound tissue, you do not have to worry about keeping a tree alive, and you can just saw away, which makes the sampling much easier. Transporting fire-scar samples out of the field and back to the lab is laborious, especially when compared with carrying a bundle of tree cores in your backpack. It involves a lot of schlepping of heavy wood wedges and cookies and then a lot of logistical headaches when shipping the hundreds of pounds of wood across or between continents.

Tom Swetnam, the former director of the LTRR, is the scientist who put tree-ring-based fire-history research on the map. I first met Tom at a welcome BBQ he threw for me at his home after I accepted his job offer at UA. During the after-party cleanup, Tom blasted Neil Young on full volume throughout the entire house. I immediately liked him for that. Tom, his main longtime accomplice in research Chris Baisan, and their team started collecting fire scars in the American Southwest in the late 1970s. They have since developed a network of fire-scar chronologies that includes more than 900 sites and extends over much of western North America. Many of those fire-scar chronologies have now been compiled and made publicly available in the International Multiproxy Paleofire Database, a database not unlike, but separate from, the International Tree-Ring Data Bank. The oldest record in the database is based on giant sequoia groves in

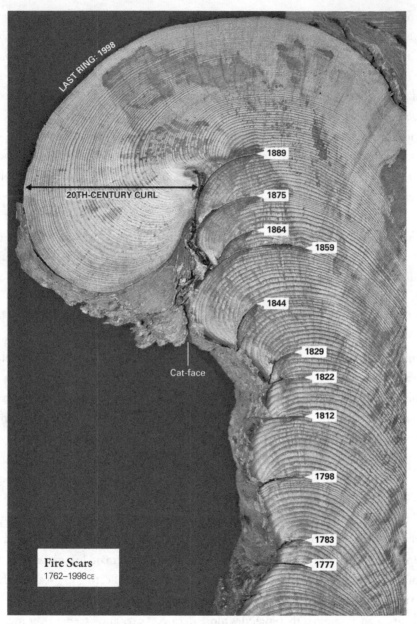

Figure 23 Wildfires can leave scars in trees. By crossdating the tree-ring sequence of this Jeffrey pine from northern California against a reference tree-ring chronology, we can date each scar to the exact calendar year when the fire occurred. During the 1800s, there were frequent surface fires. The twentieth century was largely fire free as a result of fire-suppression efforts by the US Forest Service, which was established in 1905.

Yosemite and Sequoia Kings Canyon National Parks. To sample the massive sequoia stumps, Tom and Chris had to cut more than 500 partial cross sections with a chain saw to reach deep enough inside the giant cat faces. The oldest fire scar they retrieved this way dates back to 1125 BCE.

But when it comes to fire scars in the western US, it is not just the 3,000-year-old ones that are interesting; the most recent ones are too. Any fire-history researcher working in the area can confirm how difficult it is to find a single fire scar that dates to the twentieth or the twenty-first century. One reason for this is that many of the scars that dendrochronologists date come from stumps of trees that were cut in the second half of the nineteenth century, during the rush of European settlement. But even in the vast majority of living trees, the most recent scars generally date back to the late nineteenth century, followed by evidence of more than a century of tree growth undisturbed by fire (see fig. 23). This century-long period without scars is extraordinary in the context of fire history in the American West. Prior to the twentieth century, fires occurred frequently, and intervals between fires were short, as evidenced in the Dog Valley tree stump mentioned above, with 33 fire scars in 300 years, signaling a fire roughly every 10 years.

The mystery of the missing twentieth-century fire scars is rather easily solved by looking at the history of US forest management. In 1905, President Theodore Roosevelt Jr. established the US Forest Service, charging it with conserving, protecting, and managing 300,000 square miles of National Forest land. Most of that land lay west of the Mississippi River, and Roosevelt's conservation tactics initially ran into stark opposition from western timber, railroad, and mining companies, who wanted unlimited access to the land. But the resistance to Roosevelt's forest-protection plans came to an abrupt halt after the Great Fire of 1910. In two days, the Great Fire burned more than 4,700 square miles* of forest in the northern Rockies and killed more than 100 people, many of whom were Forest Service firefighters. Following the Great Fire of 1910, the Forest Service rangers who fought it became national heroes, and fighting forest fires became a national cause.

The Forest Service has been so successful at fighting wildfires over the past century that it is very hard to find any evidence at all of twentieth-century

---

*Approximately the size of the state of Connecticut.

wildfires. The abrupt transition from frequent wildfires in the nineteenth century to nearly zero fires in the twentieth century is known as the *Smokey Bear effect* among fire historians. This nickname is an anachronism, because the firefighting advertising mascot did not enter the fire-prevention scene until the 1940s. With the popular slogan "Only YOU can prevent forest fires," Smokey Bear appeared on billboards, radio shows, and cartoons, becoming a symbol of how damaging forest fires can be and insisting that they should be prevented and fought at all costs. Tom Swetnam has been collecting original Smokey Bear posters for decades and used them to decorate the LTRR's fire-ecology lab. When you enter the lab, anywhere you look, Smokey Bear points an ironic finger at you.

Through decades of fire-history work across the American West, we have learned that frequent surface fires actually are not bad for the dry forests that can be found there. On the contrary, they are needed to keep the forests healthy and vital and to keep undergrowth fuel conditions in the forests at bay. On the other hand, a century-long lack of fire, a *fire deficit*, during which blazes have been systematically suppressed, has enabled undergrowth to become unnaturally dense. The dramatic change in fuel density and structure resulting from the twentieth-century Smokey Bear effect has turned many western fire regimes from low-intensity surface fires into high-intensity, high-severity, stand-replacing crown fires. We are now experiencing the dangerous effects of a century of intense firefighting in forests where fires should not have been doused. The upward trend in large, destructive wildfires in the American West over the past four decades can be attributed at least in part to a century of "zero-tolerance" fire-management practices that resulted in an absence of necessary, frequent surface fires.

But there is more to the story than that. The November 2018 Camp Fire and the December 2017 Thomas Fire, for instance, demonstrate the extended length of the fire season in twenty-first-century California. Whereas fires historically occurred primarily during the hot, dry Californian summer and fall and were largely absent during the wet winters of its Mediterranean climate, they now occur year-round. The prolonged fire season suggests that our current wildfire predicament is not just the result of a century of fire suppression and that merely "raking the forest" will not suffice to fix it. The shift to more high-intensity fires is happening in sync with globally warming temperatures and increasing drought. Warmer temperatures lead to earlier snowmelt and longer fire seasons. They also result in hotter droughts, which make fuels more combustible. Long, severe droughts also kill trees and thus add even more fuel to the already above-average *fuel load* of western forests. Getting us out of this predicament will require increased federal funding, not just for more robust fire suppression and protection of structures and people but also for fuel reduction through controlled burns, or intentionally set wildfires, and thinning to reduce forest combustibility.

The lack of fires in the twentieth century due to the Smokey Bear effect makes it difficult to study fire in the western US. If it weren't for fire-scar chronologies, we would not even know that the past "normal" for western dry forests was a frequent, low-intensity fire regime. Without a detailed fire history that spans the centuries predating Smokey Bear, it is also difficult to study the influence of climate on fire, to disentangle the effects of fire suppression from the effects of anthropogenic climate change on our current wildfire situation, and to project what will happen in the coming decades as temperatures continue to rise. Fortunately, the fire-scar network in the American West, which is spatially explicit and covers many centuries, is just what we need. We can, for instance, compare dendrochronological fire-history records with fully independent drought reconstructions, such as the North American Drought Atlas, to see that drought has always been a strong driver of wildfire activity in the West. Whether you are in the Southwest, the Rockies, California, or the Pacific Northwest, over the past 500 years most large wildfires have occurred in years with the driest summers. The years in our fire-history network when many trees were scarred thus reflect dry forest conditions that are conducive to large fires, rather than increased frequency of ignition sources, such as lightning. No matter how many matches (as an example of an ignition source) you throw into a wet forest on a wet day, they likely will not cause much trouble. If you light a single match on a dry day in a dry summer, however, a large wildfire is likely to ensue. It comes as no surprise, then, that the western US fire-history record consistently reflects dry years. But the picture is more complicated than that.

In the American Southwest, most years are dry, but not all years result in widespread wildfires. A dry spring and summer alone does not necessarily create a big fire year in the dry southwestern forests: you also need fuel. And to promote fuel buildup, the years leading up to a big fire year actually need to be wetter than average, not drier. Through his fire-history research, Tom Swetnam found that the Southwest's biggest fire years in the past occurred when an abnormally dry year followed two or three abnormally wet years. These are exactly the conditions that are created by ENSO in the American Southwest: El Niño years, which are wetter than normal and create fuel buildup, are followed by their dry La Niña counterparts, during which fires occur.

Tom first saw the fire-ENSO connection while he was splitting a six-pack with his friend Julio Betancourt, a paleoecologist and paleoclimatologist at the US Geological Survey, on Julio's front porch.* Julio was studying past ENSO activity at the time, and when Tom started listing the epic fire years in his southwestern fire-scar record—1893, 1879, 1870, 1861, 1851, and so forth—Julio recognized many of them as La Niña years from his work. Tom and Julio knew they were on to something when Tom then listed years with few or no fire scars—1891, 1877, 1869, 1846—and Julio recognized those as El Niño years. They had just discovered that the ENSO system in the tropical Pacific Ocean was an important driver of wildfire dynamics in the southwestern US, a milestone in our understanding of past fire-climate interactions, which they wrote up in a paper published in *Science* in 1990.

By the time I started working on fire history fifteen years later, the Swetnam and Betancourt paper was legendary. It was the giant on whose shoulders other projects stood, including a project I worked on designed by Alan Taylor, at Penn State University, to find the climatic drivers of past wildfire regimes in the Californian Sierra Nevada. As a California native, Alan knew that ENSO does not have a strong influence on climate in the Californian Sierra Nevada. But if not ENSO, then what drove past Sierra Nevada fire variability? That is what we planned to find out. My first task on the project was to go on an eight-week fire-scar sampling spree throughout the Sierra Nevada. After a week-long introduction, with Alan showing me how to find and sample fire-scarred trees and stumps, it was time for this urban chick straight out of a crowded and sophisticated European capital to spread her wings and fly into the big, wide open. I rented a car, stocked up on groceries and bear spray, and spent the rest of the summer traveling solo up and down the Sierra Nevada, sampling in the

---

*There are some parallels here with how we first came up with the idea to use shipwrecks to study past hurricanes.

National Forests, and sleeping in Forest Service barracks. Like us, the Forest Service was very interested in learning more about the fire history of their forests, and in each location firefighting crews were ready to help me out.

It was a dream for me to spend my first summer in the US wandering the Sierra Nevada forests, getting acquainted with "wilderness," a concept that simply does not exist in densely populated Belgium. The fire crews had their fair share of good-natured fun with the inexperienced academic from Europe—"Denmark? Bulgaria? Where are you from again?" One night, I was sharing the fire barracks deep in the woods of Plumas National Forest with two firefighters. After dinner, they invited me to watch a DVD they had found lying around. It turned out to be *The Blair Witch Project*, a horror movie about three amateur filmmakers who go on a hike deep in the woods and disappear. For those who have not had the dubious pleasure of viewing the film, I will just say that it is not the kind of movie you want to watch deep in the woods with no one else around. Aside from enjoying the jovial company, teaming up with the fire crews on call was a very efficient strategy. Upon arriving in a new forest, I would spend a couple of days scouting for the "best" stumps, those with the most and best-preserved fire scars in their cat faces. That was how I found the 33-er in Dog Valley. I would then come back the next day with the firefighting crew—often backed up by a fire engine and all—who did the chainsawing to collect the samples. This was a win-win situation: I had a skilled crew to help me saw without ruining the samples,[*] and the fire crew got useful chainsaw training[†] while they were on call, waiting for the next fire to happen.

I collected more than 300 fire-scar samples that summer from 29 sites that spanned the 500 miles from Lassen National Forest in the north to Sequoia National Forest in the south. After crossdating and analyzing the samples, I combined them with the vast back catalog of fire-scar samples collected by Alan, his students, and his collaborators. We now had a data set of almost 2,000 samples from the Sierra Nevada that gave us almost 20,000 dated scars to study in our quest to understand the influence of climate on California's fire history. What this massive data set told us was that past wildfires in the Sierra Nevada had been driven by . . . drought. We had collected 2,000 samples, I had

---

[*] Delicate fire scars can be easily obliterated by an inexperienced chainsaw wielder.

[†] A useful skill when digging a fireline to control fire spread by removing all flammable material (including tree stems) from a perimeter around the fire.

spent two years analyzing 20,000 fire scars, and what we found was that it burned in the Sierra Nevada when it was dry. Duh. I'd bet we could have figured that out with a slightly less exhaustive data set.

It took us a while to realize what additional information we could extract from our Sierra Nevada–wide fire reconstruction, which spanned the three centuries from 1600 to 1907, apart from its link to drought. Eventually we did. When we merged our fire-history record with a time series of annual area-burned data for the twentieth century, we noticed three distinct shifts in the fire history. At three distinct points in time—1776, 1865, and 1904—the characteristics of the fire record had changed drastically (fig. 24). For the first 175 years of our record, fire activity was fairly stable: on average, 22 percent of the sites in our network burned each year. Then in 1776 there was a shift, and for nearly a century the Sierra Nevada fire regime became more intense, with more frequent, larger, more widespread fires. Between 1776 and 1865, 38 percent of the sites burned in any given year. After 1865, the fire regime reverted back to moderate fire activity, with on average 20 percent of the sites burned, similar to in the earliest period. After 1904, Sierra Nevada fire activity dropped to its lowest level over the entire 400-year record. None of these shifts occurred

Four Fire Regimes in the Sierra Nevada
1600–present

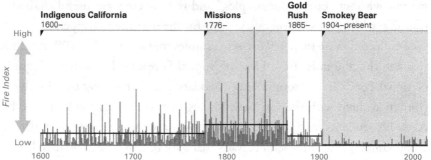

Figure 24 When we merged three centuries of fire-scar data from California's Sierra Nevada with annual area-burned data for the twentieth century, we noticed four distinct fire-regime periods. Prior to 1776, California's indigenous people used small-patch burning for agriculture and hunting. As Europeans settled North America and disease decimated the indigenous population, the frequency of large, widespread fires increased. With the influx of livestock to California after the Gold Rush, fire activity decreased again. Finally, after 1904, widespread fire suppression brought fire activity to unprecedented lows.

in any of the climate reconstructions we were working with—temperature, drought, or ENSO—which left us wondering, If not climate, then what caused the 1776, 1865, and 1904 fire regime shifts?

The most recent of the three shifts, around 1904, was fairly straightforward to explain: it coincided with Roosevelt's installation of the Forest Service and the establishment of a fire-suppression policy on federal forest lands. We also did not have to look very far to explain the prior shift, around 1865, from very high fire activity to moderate fire activity. Many of our Sierra Nevada firescar samples came from stumps of trees that had been cut around that time— the era of the Gold Rush in California. In the decade following the discovery of gold in Coloma in 1848, an estimated 300,000 people immigrated to California from the rest of the US and abroad. To meet the demands of the droves of prospectors, the state organized a rapid influx of goods and livestock. By 1862 there were 3 million sheep in California; the number had doubled by 1876. Every summer in the second half of the nineteenth century, large herds of sheep were shepherded through the forests of the Sierra Nevada to its alpine meadows, where they grazed away the fuel for potential future fires. Increased livestock grazing thus caused fragmentation and interruption of the fuel continuity, which could explain the 1865 fire-regime shift from high to moderate fire activity, from frequent large, widespread fires prior to 1865 to less frequent large wildfires after. The intensive logging needed to provide timber for mines, housing, and railroads for the Gold Rush also contributed to the 1865 fire-regime shift. We found stumps from this era even in the most remote corners of the Sierra Nevada, suggesting that logging was widespread and not just limited to towns and transportation corridors.

Both the 1904 Smokey Bear shift and the 1865 Gold Rush shift were related to anthropogenic land-use changes, which led us to look into land-use change as a possible explanation for the first shift in 1776. Up until 1776, Sierra Nevada fire activity was moderate and fairly constant through time. After 1776, fires in the Sierra Nevada became more frequent and more synchronous between sites, and the number of sites burned per year almost doubled, from 22 percent to 38 percent. In the 90 years between 1776 and 1865, there was not a single year when none of the 29 sites I had sampled in the Sierra Nevada burned. The biggest fire year was 1829, when 25 of the 29 sites—from Sequoia Kings Canyon in the south all the way to Lassen National Park in the north—

were on fire. But what happened in or around 1776 that all of a sudden made Sierra Nevada fires much more frequent and more synchronized? When I give talks about our study and ask the audience what happened in 1776, the first answer I often receive is the Declaration of Independence, which indeed happened in that year but had very little to do with California. Occasionally, audience members who are more familiar with California history may make the link to the establishment of the Spanish missions around that time, which is a connection that does bear examination.

In the years 1769 to 1833, Franciscan priests established 21 religious missions in California, aimed at evangelizing to the state's indigenous population, which comprised 500 to 600 tribes at the time of contact. The Spanish missionaries brought to California not only the Bible but also a slew of European diseases, such as smallpox, to which indigenous people had no resistance. The transmission of these diseases started almost immediately following mission establishment, and even though the missions were built exclusively on the California Coast, extensive intertribal trade networks allowed diseases to spread swiftly to the Central Valley and the Sierra Nevada foothills via trails and trade routes. *By 1855, 85 percent of the California indigenous population had succumbed to the pandemic.* Such a massive and rapid depopulation profoundly impacted the Californian landscape, including its fire regime. Prior to the establishment of the missions, indigenous tribes used sophisticated burning practices to enhance the productivity of wild tree crops, grasses, and game. The Western Mono tribe, for instance, burned the understory of blue oak woodlands to trigger the growth of Hall's mule ears (*Wyethia elata*), a perennial herb with edible seeds. After witnessing fires set by Sierran tribes, John Muir wrote in his diary in 1894 that "Indians burn off the underbrush in certain localities to facilitate deer-hunting." The California indigenous people ignited many small-scale fires that reduced fuels and fuel continuity across the forest and created a burn-patch mosaic landscape. Their small-patch burning functioned in the same way that firebreaks do today: to keep fires from spreading. As the post-1769 California indigenous-population numbers spiraled downward, fire management declined and small-patch burning became less frequent. Fuels came to be more continuously distributed, and the potential for large fires to roam across the Sierra Nevada forest increased. Tom Swetnam and his colleagues found a very similar effect of indigenous depopulation on the fire regime in the Jemez Mountains, in northern New Mexico. Spanish missions were established much earlier in this region, starting in 1598. Two decades later, the expansion of the missions resulted in a large-scale depopulation of the Jemez people and, as in the Sierra Nevada, a fire-regime shift to more frequent large and synchronous fires.

Fires occurred when it was hot and dry in the Sierra Nevada throughout all four fire-regime periods—the California Indigenous period (1600–1775), the Missions period (1776–1865), the Gold Rush period (1866–1903), and the Smokey Bear period (1904–present). But this fire-drought relationship was buffered or amplified by human land use. The influence of drought on fires was strongest during the Missions period, after the decline of indigenous small-patch burning and before the onset of fuel fragmentation during the Gold Rush. During this almost century-long period following the establishment of missions, the Sierra Nevada forest was largely untouched. Without the burn-patch mosaic created by either indigenous fire management or grazing, the climate had free rein on where and when the forest burned. I believe there is a lesson to be learned from this regarding how to handle our current western-wildfire conundrum. There is no doubt that the climate in California will get hotter and drier in the coming century, which will contribute to increasingly severe wildfires. But our results show that in the past, land-use changes have modulated the relationship between fire and climate. If we manage to create a more mosaic-like forest landscape in the Sierra Nevada by reducing fuel continuity through thinning or controlled burns, then we might be able to reduce the forest's fuel load. By doing so, one hopes, we can curtail the intensity and size of future fires, as well as how seriously they might be impacted by projected climate change.

▨ Thanks to the dense fire-scar network in the American West, we have been able to put together many of the pieces of a complex, nuanced fire-history puzzle. For many other regions across the globe, our understanding of fire history is much more limited and only just starting to emerge. In 2010, after almost three decades of using tree rings to reconstruct historical fire regimes in the American West, Tom Swetnam received NASA funding to study fire history in the taiga forest of Siberia. When I arrived at the LTRR a few months later, Tom asked me to join his field campaign to Yakutia. Until then, I had only ever heard of Yakutia as a territory to conquer in the board game Risk. I had certainly never expected to visit the remote Siberian region that loomed on the northeastern part of the board, just west of Kamchatka. I did not think long before agreeing to join the field campaign, in keeping with my motto "I'll try anything once." After Yakutia, I've been more careful about applying that motto.

Yakutia, also named the Sakha Republic of Russia, is vast by all standards. It covers more than 1 million square miles—about the size of India—and yet it is home to fewer than 1 million people.[*] Tom had assembled a team of five LTRR researchers and five of our Russian counterparts for the Yakutia field campaign to collect fire-scar samples. We had planned a 10-day trip to drive the 450 miles from Yakutsk, the Sakha capital, on the Lena River, to Botulu, the hometown of our Russian collaborator, Yegor, and to sample on the drive back. We knew that we could cover only a small sliver of the Siberian vastness during our 10-day field campaign, but we did not realize that even traversing the 450-mile transect that we had selected would prove to be a very ambitious goal.

The field campaign started with a flight of 30-plus hours from Tucson to Yakutsk. Fortunately, we had a couple of days of downtime in Yakutsk to buy last-minute supplies and to visit the local mammoth museum before heading out on the road. It was hot and muggy in Yakutsk, 95 degrees with at least 70 percent humidity, and it would stay that way throughout our trip. The two field vehicles that rolled up in front of our hotel when it was time to depart were not Jeeps or SUVs but Soviet-era vans with only primordial four-wheel drive and no air conditioning. If you look up directions in Google Maps for travel between Yakutsk and Botulu, which boasts a population of 815, you get the following message: "Sorry, your search appears to be outside our current coverage area for driving." That sounds about right. The road from Yakutsk to Botulu is straight, unpaved, and bordered by dense larch and pine forest on either side. It is primarily used in winter, when it is frozen and drivable. We discovered firsthand that in summer the road resembles a giant mud bath in which deep ruts compete for the driver's attention with pools of muck three to four feet deep.[†] It took us four days—without sampling stops—just to reach Botulu from Yakutsk. Most of that time was spent outside of the vans to (1) re-

---

[*] According to the 2010 census. At the same time, India had more than 1.3 billion inhabitants.

[†] Tyson Swetnam, one of the LTRR field-crew members, posted a series of videos on Youtube documenting our Yakutia adventure. They might be the best way to get a real feel for what it was like: https://www.youtube.com/watch?v=9n_fElk6mTo.

duce the weight of the vans so that they would not sink in the mud pools or get stuck in the ruts, (2) push the vans out of the mire, or (3) wait for our drivers to build impromptu wooden platforms to cover the deepest pools before attempting to cross them. Inevitably, one of the vans also broke down, which added an additional two days to our journey. While we waited in a nearby village for a spare part to arrive, Yegor, our Russian expedition leader, tried to cheer us up by repeating his mantra "There used to be no roads, only directions." To be honest, after four muddy, sweaty days, I started to think that "no roads" might have been better than the road we were on.

Taking the long slog to Botulu into account, it is remarkable that we still managed to sample more than 300 trees along that 450-mile stretch. The excruciatingly long workdays certainly contributed to our sampling productivity. In July, at 62 degrees north, we had about 19 hours of daylight to work with, plus a couple of hours of twilight on either side of dusk and dawn. The latest time we ever sampled a site, as recorded by the clock on our GPS, was 11:54 p.m. I don't recall seeing the night sky or even faint stars for those ten nights. On a typical day we would get up around 6:00 a.m., have breakfast, pack up our camp, and start driving (or, rather, pushing, waiting, and walking) and sampling. We'd take a lunch break from about 4:00 to 6:00 p.m. and finally stop driving around midnight, at which point it was still light outside. After pitching our tents, eating dinner, and partaking of the obligatory vodka shots, we'd go to sleep around 2:00 or 3:00 a.m.—before the whole process started again at 6:00 a.m. the next day. Our Russian collaborators were incredibly gracious in all of this: they woke up before us to have coffee ready, prepared our meals, and then stayed up after us to clean up after dinner. I wonder if they slept at all. In addition, they had brought along a camp shower tent, which they meticulously pitched every night and for which they even heated up a few gallons of water. Quite the luxury in the Siberian taiga, but I suspect that I, as the only woman on the team, was the only one who made time in our brutal schedule to enjoy it. Despite these small bouts of comfort, I was suffering. Long after we'd safely made it back to Tucson, I was glad to hear Tom say that the Yakutia field campaign had been the most difficult one of his entire career. If a hardened dendrochronologist with more than three decades of fieldwork under his belt had found the Yakutia campaign challenging, then my own discomfort was probably validated.

It was the combined effect of a long list of issues that made the Yakutia field campaign so exhausting: the heat; the ubiquitous mosquitoes and horseflies; the loud rattling noise of the vans, which made conversation impossible; the mud; the frustratingly slow progress; the endless workdays; and being the only woman on a team alongside nine men, none of whom apparently considered the benefits of showering during 10 sweaty days of hard physical labor and camping in the wild. But the one factor that trumped all others was my hypoglycemia. When the sugar level in my blood is low, I don't function well, and I get hangry, which is not pleasant in the best of circumstances. On the Yakutia trip, our Russian colleagues were in charge of everything relating to meals: they had done the shopping, they prepared the food, they decided what we ate and when we ate it. The food they prepared was delicious, but the meal schedule was brutal. After breakfast at 6:00 a.m., we ate nothing until "lunch" at 4:00 p.m., and then we didn't eat again until "dinner" after midnight. Eight hours between meals does not jive well with hypoglycemia. In addition, those long, eight-hour stretches between meals were chock full of dragging cars out of the mud and walking for miles through the forest to core trees and chainsaw logs. After evaluating the situation on the first day and realizing—with my earlier experience in the Pyrenees in the back of my mind—that my male colleagues would rather starve than admit to being hungry, around noon on the second day I asked for a snack. With some reluctance, Yegor produced an apple for each of us. When I tried the same thing the next day, we were out of apples. We had packed exactly ten apples for ten people on a ten-day trip.

Fortunately, as an experienced hypoglycemic fieldworker, I had come prepared. I had stocked up on prepackaged, chocolate-covered Belgian waffles during my stopover in Brussels and packed 10 of them in my luggage. When we ran out of apples, with my blood-sugar level crashing and no relief in sight, I tore into the first waffle while driving in the van. I couldn't wait to get that three-by-three-inch square of chocolate-covered heaven into my bloodstream. But I made the mistake of looking up. Five pairs of voracious eyes were glued to the waffle in my hand. I had no choice but to share my measly waffle with five other desperate souls. Having learned a harsh lesson from the apple debacle, I rationed our waffle stock: for the rest of the trip we shared one waffle a day among the six of us in the van. Not quite enough to keep hypoglycemia at bay, but I was impressed by what Belgian waffles and chocolate can do for morale. To this day, more than six years later, one of my Yakutia field companions regularly sends me emails praising the many virtues of Belgian waffles.

We found fire-scarred trees everywhere we looked in the Yakutia forest, and we ended up sampling 32 sites along the Yakutia-Botulu transect. The cat-faced pines and larches we sampled were 200 to 300 years old and showed anywhere from 2 to 16 fire scars. The oldest fire scar we found on a downed log dates back to 1304. The youngest scar was the result of a fire in 2010. Combining the fire-scar dates for trees from each individual site gave us a good idea of the site's *fire return interval*, that is, how frequently it burned. By lining up the fire scars from all sites combined, we figured out how frequently the area as a whole burned and which past years were the big fire years. We are thus slowly building up a network of fire-scar sites throughout Siberia that covers more and more ground. Through collaboration, the network now also extends into Mongolia and northeastern China. But the growth of this Asian fire-history network is slow and will require many more exhausting field trips. It will be a while before the network will enable us to put current wildfire dynamics in this remote corner of the world into a historical perspective as we have done for the western US.

Sixteen
The Forest for the Trees

In 1995, after 12 years of dogged excavation in sediment layers exposed by the open pit of the Schöningen coal mine in eastern Germany, the archeologist Hartmut Thieme and his team* stare down at four undeniably ancient wooden spears emerging from the thick mud.

Three of the carefully constructed and purposefully balanced spears looked to be javelins around 6 feet long, while one shorter spear with two sharpened ends was likely used for thrusting. The spears were accompanied in this particular mud layer by bone fragments from more than 20 horses showing clear signs that they had been butchered. As a former lakeside area, the Schöningen site is waterlogged; therefore, organic materials—such as the wooden spears and the horse bones—are remarkably well preserved.

The "Horse Butchery Site," or the "Spear Horizon," as it came to be known, appeared to be such a stunning archeological find that Thieme invited about 25 of his colleagues to visit and observe the site with their own eyes. After his trip out to the mine in 1995, the American prehistorian Nicholas Conard described the visit as follows: "Thieme's claims to have discovered dozens of butchered large Mosbach horses, several perfectly preserved wooden spears, multiple fireplaces, and numerous lithic artifacts in close archaeological association surpassed what any sane archaeologist could reasonably consider to be possible. . . . On November 1, I made the long train ride from Tübingen to Schöningen, without ever considering that Thieme's claims could possibly be true. . . . The mood on this day can only be described as euphoric, as all present

---

*From the Niedersächsische Landesamt für Denkmalpflege, the Lower Saxony Heritage Office.

quickly realized that they were witnessing a discovery unrivaled in the annals of archaeology."*

While the organic material at the site was too old for radiocarbon dating, or for tree-ring dating for that matter,† by examining the layers underneath and above the site using other techniques, such as thermoluminescence dating,‡ it was possible to calculate a fairly precise date range for the wooden spears. As it turns out, the spears are between 337,000 and 300,000 years old, making them the oldest wooden artifacts in human history. They predate even the rise of the Neanderthals nearly 300,000 years ago and were likely made by *Homo heidelbergensis* hominins, a species of archaic human that displays traits of both *Homo sapiens* and our more distant progenitors, *Homo erectus*, who roamed the earth as early as 1 million years ago. The Schöningen excavations led to a seismic paradigm shift in our understanding of hominin behavior and human evolution in the early Stone Age. The Paleolithic wooden spears indicate that the Schöningen *Homo heidelbergensis* used sophisticated weapons and tools and were skilled hunters at the top of the food chain. This required a level of planning, social coordination, and communication that previously typically was attributed only to modern humans, not pre-Neanderthal hominins.

The Schöningen spears also powerfully demonstrate that sophisticated wood use has been part of human culture since the early days of human evolution, more than a quarter million years back in the distant past. This makes sense. Early humans would have used wood as a resource because it was widely available, easily accessible, and required no sophisticated tools for processing. Throughout time, wood has allowed humans to cover their basic food, shelter, and energy needs. Over millennia, as more durable copper, bronze, and iron tools for woodworking replaced stone axes, wood craftsmanship advanced steadily and wood applications became ubiquitous. In turn, this ample prehistoric and historic wood use has allowed dendroarcheologists to precisely date and analyze archeological finds across the globe.

*N. J. Conard, J. Serangeli, U. Böhner, B. M. Starkovich, C. E. Miller, B. Urban, and T. Van Kolfschoten, "Excavations at Schöningen and paradigm shifts in human evolution," *Journal of Human Evolution* 89 (2015), 1–17.

† That is, it was more than 50,000 years old.

‡ Some materials, such as flint fragments from Schöningen, have accumulated energy over a long period of time. When they are pretreated and heated, they become luminescent. The amount of luminescence changes with the age of the material and can be used to date the material.

Fortunately for dendrochronologists and students of history, wood use in construction has a longer timeline and has been even more widespread and diverse than intuition would indicate. The earliest evidence of a man-made wooden construction dates back to ca. 9000 BCE, the middle Stone Age or Mesolithic, and was found at Star Carr in North Yorkshire in the United Kingdom. Star Carr archeologists did not actually find a wooden construction but rather 18 postholes arranged in a circle 13 feet wide, indicating that a circular wooden hut once stood at the site. As the Star Carr postholes exemplify, little remains of ancient aboveground wooden constructions, which gives the impression that past civilizations were less reliant on wood than they actually were. The Romans, for instance, made ample use of wood in their constructions, to create molds for bricks, for instance, or to build timber cranes to create formidable structures appropriate for decadent Roman emperors. But of the many Roman wooden constructions, only the wood that was used to line water wells and has been waterlogged since remains. Such waterlogged environments, where the wood is preserved in anoxic conditions, give us a glimpse of wood use and carpentry history. Waterlogging gave us the Paleolithic Schöningen spears, the Neolithic Lake Murten pile dwellings, and the English Sweet Track and taught us that the first farmers in Europe, in the sixth millennium BCE, were also the first carpenters. Wood in more recent constructions, such as the medieval Gothic cathedrals in Europe or the Ancestral Puebloan great houses and kivas, has had less time to decay, can be found above ground, and can often be tree-ring dated.

In more recent centuries, wood was the pivotal power behind the Age of Discovery and the Industrial Revolution. The Spanish shipwrecks we tallied in our reconstruction of Caribbean hurricanes were built from hardwood. The scarred stumps we used to reconstruct fire history in the Sierra Nevada were mostly from trees cut during the mining boom of the Comstock era (1859–74). The silver rush following Henry Comstock's discovery of prized ore on the eastern slopes of the Sierra Nevada in 1859 demanded vast quantities of wood to build mines, mining camps, and mills, as well as the wagons and railroads needed to transport ore and supplies to and from the silver mines. Wood was needed throughout the world not only to construct mines but also to smelt the metals derived from them at high temperatures.

The many archeological and historical wood remnants—buildings, water wells, artifacts, charcoal, tree stumps—represent only a fraction of the ways in which we have leveraged this unparalleled natural resource. Wood has been used to make weapons for hunting and warfare, to make tools, furniture, sports equipment, woodblock printing, and paper, even making possible the written word as you see it in this very book today. Wood was the primary source of energy both in homes and in industry up until the Industrial Revolution and the pervasion of fossil fuels. It is no exaggeration to say that human civilization as we know it is founded on trees.

In 1774, Captain James Cook lands his Royal Society–funded vessel, the HMS Resolution, *on a sandy beach on a remote island in the South Pacific Ocean, more than 2,100 miles west of South America. As he approaches the island, a few "upright standing pillars" come into view in an otherwise bleak and bare landscape.*

The long history of human wood use and the deforestation that it necessitated have left their mark on landscapes, human societies, and the earth system as a whole. One of the clearest examples of extreme deforestation can be found at Rapa Nui (Easter Island), one of the last inhabitable islands to be settled by humans. We know from pollen data extracted from volcanic crater sediment cores that when chief Hotu Matu'a first set foot on Easter Island around 1200, the island was forested with giant palms and about twenty tree species. By the time the first European, the Dutch explorer Jacob Roggeveen, reached the island in 1722, not a single tree remained. Over the 500 years since their settlement of the island, the Rapa Nui people had completely denuded the island of its forests, extinguishing all of its native tree species. If you visit the island now, all you will find in terms of natural vegetation is grassland and the occasional bush. But it is unlikely that you would fly all the way to the most remote island on earth to see its natural vegetation. More likely, it would be Rapa Nui's giant statues—the Moai—that attracted you. More than 900 Maoi were carved from volcanic tuff (rock made of compacted ash) by the Rapa Nui people between ca. 1400 and 1680. Nine hundred megalithic statues, the largest standing more than 30 feet high and weighing more than 80 tons, is a lot for an island with an area of less than 165 square miles. To transport and erect all these Maoi, the Rapa Nui people needed large quantities of wood and rope made from tree bark. They further needed timber to build oceanworthy canoes and for housing and heating. For all of these purposes, they started clearing Rapa Nui's forests soon after their arrival, and deforestation reached its peak in the 1400s. By the 1600s, Rapa Nui's landscape was the wasteland that Jacob Roggeveen and Captain Cook witnessed in the eighteenth century and that it still is today.

In Chapter 8 we saw how, prior to Rapa Nui's deforestation, the same fate befell Iceland, on the other side of the globe. In 874, when the Norse arrived, one quarter of Iceland was covered by forest. But in little under three centuries, the Norse settlers removed virtually all of the island's forests for fuel, timber, and agricultural use. From very early on, the Norse settlers had to import wood from the Scandinavian mainland. The National Museum of Iceland exhibits a vast collection of wooden artifacts reflecting Iceland's Norse history, but with the exception of one thirteenth-century crucifix, all objects are made of imported wood. At the start of the twenty-first century, only about 1 percent of the island is covered by forest despite intensive forest-restoration efforts. Iceland therefore is a blank spot on the global tree-ring map.

Not all episodes of forest clearance in history were as extreme as those of Iceland and Rapa Nui, where virtually all native forest was removed, never to return. But the history of wood depletion and deforestation is long. In fact, deforestation was mentioned in the oldest known surviving work of literature in human history, the third-millennium BCE *Epic of Gilgamesh*, which includes a story titled "The Forest Journey." Gilgamesh was the king of Uruk, a Mesopotamian city-state in present-day Iraq. To ensure that his name would endure throughout history, Gilgamesh decided to build a temple, a palace, and city walls, which required vast amounts of wood. Fortunately for him, 5,000 years ago the mountain slopes of Mesopotamia had vast cedar forests. In "The Forest Journey," Gilgamesh sets off on a journey to the Ancient Cedar Forest. He first fights and defeats Humbaba, the monster guard of the Cedar Forest, appointed by the Mesopotamian gods to protect it from the greed of man. Unfazed by this encounter, Gilgamesh then cuts down the entire Cedar Forest, including the sacred, tallest cedar. He builds a cedarwood raft to float down the Euphrates and uses the wood from the sacred cedar to make a gate for his city of Uruk.

Whereas the *Epic of Gilgamesh* is a mythical poem and Gilgamesh is a fictitious king, "The Forest Journey" represents the fate of the Middle Eastern cedar forests that were destroyed over millennia of human civilization and wood exploitation. Since Gilgamesh's days, the global human population has grown exponentially, and with it our demand for wood. Forests worldwide have been cut down not only to provide timber and fuel resources but also to clear land for agriculture to feed the growing world population. We find examples of such historical deforestation processes throughout the world, but they have been particularly well documented in Europe, with its long history of dense population. Italy was almost entirely stripped of its forests in Roman times, when the Roman Empire's need for timber and charcoal skyrocketed. The Iberian Peninsula followed suit in the fifteenth and sixteenth centuries, when early modern Spain developed its transatlantic empire and needed timber for its fleets of ships traveling to and from the Americas. There is a theory that the Monegros Desert in Aragon, in northeastern Spain, is the result of intense deforestation. In the 1580s, King Felipe II cleared the Aragon forests to construct the armada that attacked England in 1588. The invasion of England failed, leaving large parts of Spain deforested and without resources to maintain the country's hegemony at sea. Nowadays, where the mighty Aragon forests once stood, there is a barren landscape that teems with life only during the annual pilgrimage of electronic music fans to the Monegros Desert Festival.

Efforts to mitigate and adapt to wood depletion have also been well documented in European history. We know from dendroprovenancing that the countries surrounding the North Sea—England, France, Belgium, and the Netherlands—started importing wood from the Baltic region as early as the 1200s. From the fifteenth century on, the Venetian Republic implemented sophisticated forest-conservation policies on its mainland holdings to secure sustainable timber supplies to build its ships and to maintain its extensive levee system. Venice was way ahead of its time in keeping large forest preserves under state protection; some are still standing today. In Belgium, the successive wood demands of the Romans, the medieval cities, and commoners and nobility alike in the modern period depleted the Gallic forests, aptly named

Silva Carbonaria (Charcoal Forest), and the *Arduenna Silva*. To feed the energy demands of the nineteenth-century Industrial Revolution, large areas of southern Belgium—the Ardennes—were reforested with fast-growing spruce trees. One hundred fifty years later, these relatively young, monotonous stands constituted my concept of a forest as I was growing up. In England, the wood and energy shortages leading up to the Industrial Revolution were addressed in a different way: by using coal. Mineral (or anthracite) coal has been burned as a domestic fuel in the British Isles since medieval times, but large-scale mining started in the mid-eighteenth century, when England shifted from charcoal to mineral coal as its major energy source. English industrialists developed a refined version of coal—coke—that was clean enough to be used for iron smelting and thus for steel production. The combination of coal as an energy source and steel as a construction material set the stage for the Industrial Revolution in England, and the rest of the world soon followed suit.

As the Industrial Revolution progressed, coal was progressively supplanted by two other fossil fuels: oil and gas. Because these fossil fuels originate in organic material—plants and plankton—they contain a lot of carbon. They are the slow, million-year-scale storage component of Earth's natural carbon cycle, in which carbon is exchanged between atmosphere, land, and ocean. The land-atmosphere carbon cycle is kept in balance by two main, much faster processes: respiration and photosynthesis. Animals (including humans) and plants respire carbon dioxide into the atmosphere, which is then extracted back from the atmosphere by photosynthesizing plants. Plants use the carbon that they take from the atmosphere to grow their leaves, roots, and wood. When plants die and decay, their carbon gets incorporated into the soil, then soil microorganisms respire it back into the atmosphere. On long timescales of millions of years, some dead plant (and plankton) material also gets incorporated into the deeper layers of the earth as coal and natural gas. In the natural, balanced carbon cycle, the carbon stored in the geosphere is released into the atmosphere at equally slow rates, through weathering and metamorphism.[*]

By burning fossil fuels, we have dramatically sped up this part of the carbon cycle, pushing it out of balance. In less than 200 years since the start of industrialization, we have emitted millions of years' worth of fossil-fuel car-

---

[*] The slow change in the mineral structure of rocks due to heat, pressure, and/or chemical processes.

bon into the atmosphere. By doing so, we have thrown the natural carbon balance out of kilter, adding carbon to the atmosphere at lightning speed (relatively speaking) and thus enhancing the natural greenhouse-gas effect. The impacts of this enhanced greenhouse-gas effect are well known and already under way: rising global temperatures, melting glaciers and ice caps, rising sea levels, more heat waves, droughts, and floods, the polar vortex, longer wildfire seasons, and so forth. In fact, the impacts of the enhanced greenhouse-gas effect on the earth's system have been so far-reaching that the most recent geological epoch is now referred to as the *Anthropocene*, a period during which humans constituted the strongest driver of change to the earth's system, leaving a permanent mark in the geological record. For instance, the plastic water bottles and fake plastic trees we produce and consume in seemingly infinite numbers do not only show up as a giant garbage patch in the Pacific Ocean; they also form *plastiglomerate*, a new type of rock made up of melted plastic, sand, and basalt. If humans disappeared from the planet today, the changes we have made to the earth's atmosphere, biosphere, hydrosphere, and geosphere would still be detectible thousands of years from now.

In 1946, at the beginning of the Cold War nuclear arms race with the Soviet Union, the United States starts a nuclear testing program at Bikini Atoll, in the middle of the tropical Pacific Ocean. Over the course of the next 12 years, 23 nuclear devices are detonated at the atoll, including a thermonuclear hydrogen bomb, Castle Bravo, which by itself produced a 15 megaton explosion, about 1,000 times more powerful than the atomic bombs that destroyed Hiroshima and Nagasaki.

The anthropogenic impact on the earth accelerated drastically after World War II, when globalization, industrialization, and population growth all increased in sync, leaving a trail of fossil-fuel emissions, soot, plastic pollution, and radioactive elements in their wake. The onset of the Anthropocene is therefore often placed at the end of World War II, the start of aboveground nuclear-bomb tests so impactful that they left a permanent and traceable radioactive mark in biological and geological archives, such as tree rings and lake sediments. To demonstrate the truly global extent of the impact of the nuclear-bomb tests and of the Anthropocene, the dendrochronologist Jonathan Palmer, of the University of New South Wales, and his colleagues studied the tree rings of the "world's loneliest tree," a Sitka spruce (*Picea sitchensis*) growing more than 170 miles away from its nearest neighbor tree.* Planted in the early 1900s, it is the only tree on Campbell Island, in the Southern Ocean, far south of New Zealand. When measuring the radiocarbon content of the spruce's tree rings, Jonathan found a radiocarbon "bomb peak" in the ring of 1965, two years after the Partial Test Ban Treaty, which ushered in the end of aboveground nuclear-bomb testing. Jonathan and his colleagues thus put the onset of the Anthropocene in 1965, the year when the nuclear tests left their mark even on the loneliest, most remote tree on earth.

Other scientists, however, argue that humans have had an indelible influence on planet Earth for much longer and that our relationship with trees is at the center of it. The paleoclimatologist Bill Ruddiman, of the University of Virginia, is a strong advocate of this "Early Anthropocene" theory. In his book *Plows, Plagues, and Petroleum*, Ruddiman suggests that the irreversible impact of humans on the earth's system, and in particular on the atmosphere, started much earlier than the 1960s or even than the Industrial Revolution. Ruddiman postulates that anthropogenic impacts started 8,000 years ago, with the first agriculture and deforestation, and that they slowly grew stronger up until the Industrial Revolution, at which time their growth accelerated greatly. Deforestation leads to less photosynthesis and less carbon storage in wood; as a result, less carbon is taken out of the atmosphere than is put into it through respiration. As early farmers in southeastern Europe started clearing forest to grow crops around 6000 BCE, they started removing the photosynthetic component from the natural carbon cycle, destabilizing the cycle. The resulting carbon-dioxide overdose in the atmosphere increased steadily as agriculture and deforestation spread. Ruddiman's Early Anthropocene theory further addresses the small dips in the rising carbon-dioxide curve by laying bare the darkest aspect of the impacts of deforestation. Continental-scale pandemics such as the plagues in the sixth and fourteenth centuries and the post–European colonization smallpox epidemics in the Americas killed tens of millions

---

*For a video of Jonathan coring the tree, see https://theconversation.com/anthropocene-began-in -1965-according-to-signs-left-in-the-worlds-loneliest-tree-91993.

of people and allowed trees to recolonize hundreds of millions of acres of former farmland, thus increasing the photosynthetic capacity of the land, and with it temporarily decreasing the carbon-dioxide concentration in the atmosphere. Baldly stated, the more death visits people, the better off forests are.

The Early Anthropocene theory is strongly debated in geoscience circles, mostly because of disagreements about the scale and rate of past agricultural spread and deforestation. The impact of deforestation on the carbon cycle and related greenhouse-gas increases, however, is based on first principles and beyond debate. In the twenty-first century, tropical deforestation is responsible for about 30 percent of the enhanced greenhouse-gas effect. The good news is that the inverse is also true: by adding forest, we can add photosynthesis and take more carbon out of the atmosphere. We have seen this happening in Europe: during the modern period (ca. 1500–1850), many European forests were cleared for agriculture and for wood use, as the Aragon forests were to build the Spanish Armada. But from the beginning of the nineteenth century on, rural areas in western and central Europe and Scandinavia were abandoned as people moved to cities, where the Industrial Revolution was providing jobs, and forests resurged. Many abandoned rural areas were either encroached by surrounding forests or actively reforested, as was the case in the southern part of my homeland, Belgium. So while Europe's land lost carbon to the atmosphere through deforestation up until ca. 1800, much of that carbon loss was reversed in the following two centuries through *afforestation* (on land where there had been no forest before) and *reforestation* (on land where there had been forest). But even though Europe's forest transition might have reversed its carbon loss due to deforestation, it did not compensate for the massive amount of carbon put into the atmosphere through burning of fossil fuels.

To counterbalance the greenhouse-gas emissions from burning fossil fuels, we would need to minimize ongoing deforestation and plant new forests on a massive, global scale. There is something to be said for this strategy: forests do not pollute; they produce goods and services, such as wood and ecotourism, that can be managed sustainably. Also, it is possible that the enhanced greenhouse-gas effect itself might leverage our forestation efforts. As global temperatures rise, large stretches of land that used to be too cold for forests to grow now have potential for afforestation. Good examples of places with this potential are the North American and Russian Arctic regions, where there is ample land for new forests and where the warming has been two to three times faster than the global average. At such high latitudes, higher temperatures lengthen the growing season, which further enhances the potential for forests to take up carbon. Finally, some studies examine the potential future role of *carbon fertilization*, in which plants will photosynthesize more and capture more carbon from the atmosphere as the concentration of carbon dioxide in the atmosphere rises. Carbon fertilization functions like my dog, Roscoe: if I put a small bowl of food in front of Roscoe, he will eat it all and maintain his health; if I put a big bowl in front of him, he will still eat it all and grow ever plumper. The carbon-fertilization theory assumes that like Roscoe, trees are not very good at self-control.

Unfortunately, the solution to our man-made climate-change problem is not as simple as just planting more trees. Adding million-year-old carbon from fossil fuels to the contemporary carbon cycle and then relying on current and future forests to sequester it is a highly risky gamble. There are many caveats involved. First of all, forests need more than carbon alone to grow: they also need space and water and nutrients, such as nitrogen and phosphorus. The need for water and nutrients may reduce carbon fertilization: no matter how much extra carbon you feed the trees, if the amount of water and nitrogen stays the same, their fertilization will be limited. Furthermore, new forests compete for land, water, and nutrients with food production. We cannot plant the entire planet with trees without failing to feed the 7.5 billion people that inhabit it. Even in areas that are better suited for growing trees than for growing rice or potatoes, there may be unintended consequences to afforestation. The greening of the Arctic, for instance, will transform large areas from snowy white to dark green, thus reflecting much less incoming radiation from the sun and reinforcing radiation-caused warming. Then there are forest disturbances to consider. Decades or even centuries of carefully planned forest growth and carbon sequestration can be erased in one fell swoop when a hurricane makes landfall, when a severe drought hits, or when a wildfire rages through the forest. Diseases can spread through forests like wildfires, killing trees and their carbon-capturing potential with them. Chestnut blight, for instance, a pathogenic fungus inadvertently introduced into the US in the early twentieth century, caused a rapid and widespread die-off of the American chestnut (*Castanea dentata*), a tree species once abundant in eastern North American forests. By 1940, chestnut blight had wiped out most mature American chestnut trees on the continent. Forests can also be decimated by outbreaks of insects, such as mountain pine beetles, which attack and kill pines throughout the American West. Unfortunately, rising temperatures prolong not only the growing season of trees but the season of insect activity as well.

Given the importance of forests in the global effort to mitigate man-made climate change, we need a good grasp of the potential caveats and dangers involved in forest carbon sequestration. Countries with large land areas, such as India and China, are looking at afforestation, reforestation, and decreasing deforestation in order to comply with their commitments to international agreements, such as the 2015 Paris Climate Agreement. These agreements rely on accurate quantifications of the forest carbon-capture potential, so that land-use policies can be optimized toward carbon sequestration and climate-change mitigation. If only we had a way to look into the trees to see how much wood they have grown—and how much carbon they have stored—over time and how that growth has been influenced by water availability, climate variation, and forest disturbances . . .

Indeed, dendrochronologists have a powerful tool in hand to help solve the global carbon puzzle, quite literally. With an increment borer, we can investigate how much wood is grown and how much carbon is stored by trees of different species, of different ages, on different soils, and in different climates. We can study how lengthened growing seasons influence woody growth. How droughts, extreme weather, and rising temperatures impact growth and how these impacts change as the climate shifts. How frequently wildfires and insect outbreaks occur in a landscape and how much they impact forest growth. Tree rings have taught us how climatic changes have impacted past societies. How, when past civilizations failed, climate change was often one of the threads in a socioenvironmental web portending disintegration. How in the end the resilience of a society, rooted in its inventiveness and adaptive capacity, determines whether adverse conditions lead to a temporary regression or to total destruction.

As I write this final paragraph, man-made climate change is the main foe that we must conquer to assure humanity's prosperity. For the first time in human history, thanks to centuries of scientific discoveries and research, we have foresight into the climatic changes that lie ahead of us. Tree rings do not only help us understand how past societies have dealt with unexpected climate changes; the stories tree rings tell us, in whispers and shouts, can also inspire us to discover innovative ways to mitigate and adapt to its worst consequences. To explore this new frontier and harness its potential, dendrochronologists will need to collaborate with foresters, ecologists, geographers, sociologists, anthropologists, biogeochemists, atmospheric scientists, hydrologists, and policy-makers, among others. We have our work cut out for us.

In 2022, I assemble a team to revisit the Smolikas site in the Pindos Mountains in Greece once more. We bring the longest borer ever made and finally manage to reach Adonis's pith. As I walk down this mountain where some of Europe's oldest trees grow, I ponder their survival in a region that is defined by 3,000 years of human civilization. I am in awe of the potential for coexistence and symbiosis between people and trees. As I look out over the rugged Pindos landscape, I see windmills whirring away, generating carbon-neutral, infinitely renewable energy. I see freshly planted trees gobbling carbon dioxide out of the air with all of their young, energetic might. When we arrive in Samarina, at the foot of the mountain, I notice that since our last visit our hotel has installed solar panels on its roof. At night, we drink local red wine and toast to our credo: to do good science with good friends. Cheers!

Playlist

"The Wind Cries Mary"
Jimi Hendrix (1967)

"Once upon a Time in the West"
Ennio Morricone (1968)

"After the Gold Rush"
Neil Young (1970)

"Hey Hey, My My (Into the Black)"
Neil Young & Crazy Horse (1979)

"Africa"
Toto (1982)

"It's the End of the World as We Know It"
REM (1987)

"Disintegration"
The Cure (1989)

"Wind of Change"
The Scorpions (1990)

"Fake Plastic Trees"
Radiohead (1995)

"99 Problems"
Jay Z (2003)

"Single Ladies (Put a Ring on It)"
Beyoncé (2008)

Tree Species

Latin name	Common name	Chapter
Adansonia digitata	baobab	3
Castanea dentata	American chestnut	16
Castanea sativa	chestnut	3
Cedrus atlantica	Atlas cedar	3, 7
Chamaecyparis obtuse	Japanese cypress	5
Cryptomeria japonica	Japanese cedar	10
Eucalyptus spp.	eucalypt	3
Fitzroya cupressoides	alerce	3
Fokienia hodginsii	fokienea, Fujian cypress, po mu	12
Juniperus occidentalis	western juniper	3
Juniperus spp.	juniper	3
Lagarostrobus franklinii	Huon pine	3
Larix sibirica	Siberian larch	12
Macrolobium acaciifolium	arapari	4
Nothofagus antarctica	Antarctic beech	4
Olea europaea	olive	3
Picea abies	Norway spruce	3
Picea sitchensis	Sitka spruce	16
Pinus aristata	Rocky Mountain bristlecone pine	3
Pinus balfouriana	foxtail pine	3
Pinus cembra	stone pine	8
Pinus elliottii	slash pine	9
Pinus heldreichii	Bosnian pine	3
Pinus longaeva	bristlecone pine	3

Latin name	Common name	Chapter
Pinus ponderosa	ponderosa pine	1
Pinus sibirica	Siberian pine	12
Pinus sylvestris	Scots pine	10
Pinus uncinata	mountain pine	6
Populus deltoides	eastern cottonwood	3
Populus spp.	cottonwood	3
Populus tremuloides	quaking aspen	3
Prunus cerasus	cherry	3
Quercus douglasii	blue oak	9
Quercus petraea	sessile oak	4
Quercus robur	European oak	4
Sabina przewalskii	Qilian juniper	11
Sequoia sempervirens	coastal redwood	3
Sequoiadendron giganteum	giant sequoia	1
Taxodium distichum	bald cypress	3
Taxodium mucronatum	Montezuma bald cypress	12
Taxus baccata	yew	3
Tectona grandis	teak	4
Thuja plicata	western red cedar	10

Recommended Reading

Atwater, B. F., Musumi-Rokkaku, S., Satake, K., Tsuji, Y., Ueda, K., and Yamaguchi, D. K. 2016. *The orphan tsunami of 1700: Japanese clues to a parent earthquake in North America*. Seattle: University of Washington Press.

Baillie, M. G. L. 1995. *A slice through time: Dendrochronology and precision dating*. London: Routledge.

Bjornerud, M. 2018. *Timefulness: How thinking like a geologist can help save the world*. Princeton, NJ: Princeton University Press.

Bradley, R. S. 2011. *Global warming and political intimidation: How politicians cracked down on scientists as the earth heated up*. Amherst: University of Massachusetts Press.

DeBuys, W. 2012. *A great aridness: Climate change and the future of the American Southwest*. Oxford: Oxford University Press.

Degroot, D. S. 2014. *The frigid golden age: Experiencing climate change in the Dutch Republic, 1560–1720*. Cambridge: Cambridge University Press.

Diamond, J. 2005. *Collapse: How societies choose to fail or succeed*. New York: Viking.

Fagan, B. 2000. *The Little Ice Age*. New York: Basic Books.

Fritts, H. C. 1976. *Tree rings and climate*. London: Academic.

Harper, K. 2017. *The fate of Rome: Climate, disease, and the end of an empire*. Princeton, NJ: Princeton University Press.

Hermans, W. F. 2006. *Beyond sleep*. London: Harvill Secker.

Jahren, H. 2016. *Lab girl*. New York: Penguin Random House.

Klein, N. 2014. *This changes everything: Capitalism vs. the climate*. New York: Simon & Schuster.

Le Roy Ladurie, E. 1971. *Times of feast, times of famine: A history of climate since the year 1000*. New York: Doubleday.

Macfarlane, R. 2019. *Underland, a deep time journey*. New York: Norton.

McAnany, P. A., and Yoffee, N., eds. 2009. *Questioning collapse: Human resilience, ecological vulnerability, and the aftermath of empire*. Cambridge: Cambridge University Press.

Oreskes, N., and Conway, E. M. 2011. *Merchants of doubt: How a handful of scientists obscured the truth on issues from tobacco smoke to global warming.* New York: Bloomsbury.

Powers, R. 2018. *The Overstory.* New York: Norton.

Pyne, S. J. 1997. *Fire in America: A cultural history of wildland and rural fire.* Seattle: University of Washington Press.

Ruddiman, W. F. 2010. *Plows, plagues, and petroleum: How humans took control of climate.* Princeton, NJ: Princeton University Press.

Webb, G. E. 1983. *Tree rings and telescopes. The scientific career of A. E. Douglass.* Tucson: University of Arizona Press.

White, S. 2011. *The climate of rebellion in the early modern Ottoman Empire.* Cambridge: Cambridge University Press.

White, S. 2017. *A cold welcome: The Little Ice Age and Europe's encounter with North America.* Cambridge, MA: Harvard University Press.

Wohlleben, P. 2016. *The hidden life of trees: What they feel, how they communicate—Discoveries from a secret world.* Berkeley, CA: Greystone Books.

Glossary

Aerosol: A suspension of particles dispersed in air as a fine spray.

Afforestation: The (natural or man-made) development of new forest on land that previously has not been forested.

Age of Discovery: Early fifteenth through late eighteenth century, when European explorers traveled around the world in search of new trade routes. It marks the start of globalization and of European colonialism.

Anoxic: Depleted of oxygen.

Anthropocene: The current geological age, during which the earth system is influenced primarily by human activity. The onset of the Anthropocene is debated but is often associated with the end of World War II, when aboveground nuclear-bomb tests were so impactful that they left a permanent and traceable radioactive mark in biological and geological archives.

Anthropogenic: Man-made.

Anticyclone: A system of high atmospheric pressure associated with warm, dry weather around which air circulates in a clockwise (Northern Hemisphere) or counterclockwise (Southern Hemisphere) direction.

Buttress: Large, wide root on the side of a shallowly rooted tree that prevents the tree from falling over. Also called *butt swell*.

Cambium: A layer of living cells between the bark and the wood of trees, in which new wood and bark cells are formed.

Carbon fertilization: The phenomenon that plants photosynthesize more and capture more carbon from the atmosphere as the concentration of carbon dioxide in the atmosphere rises.

Cat face: A sequence of fire scars, left in the stem of a tree by subsequent surface fires.

Climate engineering: The deliberate and large-scale human intervention in the workings of the earth's climate system with the aim of mitigating the impacts of the enhanced greenhouse-gas effect. Also known as *climate intervention* or *geo-engineering*. See also **Solar radiation management**.

Climate reconstruction: A proxy-based, quantitative estimation of past climate variability.

Climatic determinism: An eighteenth-century approach to history that held that human activity is largely determined by climate and environment.

Clonal tree: A tree that propagates and disperses asexually through root suckers.

Controlled burn: Intentionally set wildland fire. Also called *prescribed fire.*

Cosmogenic isotopes: isotopes created by high-energy cosmic rays, such as solar flares. See also **isotopes; solar flare.**

Crossdating: The process of matching variations in tree-ring characteristics, such as ring width, among trees growing in the same climate or region to determine the exact calendar year in which each individual ring in a tree or in a piece of wood was formed.

Crown fire: High-intensity, high-severity wildfire that reaches the crowns of trees and can be destructive. Also called *stand-replacing fire* or *high-intensity fire.*

Cyclone: A system of low atmospheric pressure associated with cool, humid weather around which air circulates in a counterclockwise (Northern Hemisphere) or clockwise (Southern Hemisphere) direction.

Dendroarcheology: The dendrochronological study of wood and charcoal derived from historic buildings, archeological material, artifacts, musical instruments, and artwork.

Dendroclimatology: The study of past climate using tree-ring data.

Dendrogeomorphology: A subfield of dendrochronology that uses tree-ring data to study Earth system processes, such as erosion and glacier movements.

Dendroprovenancing: Using tree-ring data to determine the source origin of the wood used to produce an object.

Dzud: A cold winter accompanied by heavy snowfall.

Earlywood: The portion of wood in a tree ring that is formed in spring.

Earth's radiation budget: The difference between the amount of energy Earth receives from the sun and the amount it emits and reflects back into space.

ENSO (El Niño Southern Oscillation): A climate pattern that recurs every three to seven years and involves fluctuations in the ocean-water temperatures of the tropical Pacific Ocean. A complete ENSO cycle includes a warm El Niño phase, a cool La Niña phase, and an ENSO-neutral phase.

False ring: Some trees, for instance trees growing in summer monsoon climates, form more than one ring in some years. False rings can be detected by microscopically analyzing the ring boundary: the transition from a false ring boundary to the subsequent ring is gradual, whereas this transition is sharp for real annual rings.

Fire deficit: A prolonged, artificial lack of wildfire due to extensive fire suppression. See also **Smokey Bear effect.**

Fire intensity: The heat energy released by a fire. High-intensity, crown fires are hot and destructive. Low-intensity, surface fires do not reach the crowns of trees and are less destructive.

Fire return interval: The average number of years between two fires.

Floating chronology: A tree-ring chronology that has not yet been crossdated with an absolutely dated reference chronology and thus is not yet anchored in time.

Flood rings: Tree rings that occur in trees growing on riverbanks and indicate years when the trees were inundated by spring or summer floods.

Frost rings: Tree rings with irregularly shaped wood cells as a result of frost during the tree's growing season.

Fuel ladder: Vegetation that allows a forest fire to climb up from the forest floor into the tree canopy.

Fuel load: The amount of flammable material in a forest that can feed a forest fire. A small fuel load typically results in slow, low-intensity fires. A large fuel load can result in fast, destructive crown fires.

Glacial: Ice age. See also **Interglacial.**

Global climate model: Computer-run program that uses the laws of physics, fluid motion, and chemistry to mimic Earth's complex climate system. Also called *general circulation model* or *GCM*.

Global weirding: Crazy climate and weather (e.g., heat waves, droughts, hurricanes, snow storms) related to the enhanced greenhouse-gas effect and rising global temperatures.

Great house: A large, multi-storied Ancestral Puebloan structure.

Hadley circulation: Atmospheric circulation that moves warm air from the equator toward the poles.

Heritage tree: Typically a large, old, solitary tree with unique cultural or historical value.

Hindcasting: Testing a climate model by entering known past events (e.g., volcanic eruptions) into the model and comparing the model outcome with the known climate.

Holocene: The most recent geological epoch, which started ca. 11,650 years ago, after the last glaciation.

Hominin: The taxonomic tribe comprising all modern and extinct human species and our immediate ancestors but excluding other great apes, such as chimpanzees, gorillas, and orangutans.

Impostor syndrome: A psychological pattern in which an accomplished person is convinced that she or he is a fraud and experiences a persistent fear of being exposed as such.

Increment borer: A specialized tool used to extract a core from a living tree or wooden beam without injuring the tree.

Instrumental climate record: The meteorological data derived from daily measurements at weather stations around the world.

Interglacial: A geological interval, lasting thousands of years, of warmer, milder climate between two glacials. See also **Glacial.**

Isotopes: Multiple forms of the same chemical element that differ in their relative atomic mass but not in chemical properties. Isotopes can be stable (e.g., ^{12}C, ^{13}C) or radioactive (e.g., ^{14}C). See also **Cosmogenic isotopes.**

Jet streams: The fast flowing, meandering westerly winds that encircle Earth near the tropopause, at ca. 6 miles above Earth's surface. There typically are two or three jet streams in each hemisphere.

Kiva: A large, often subterranean room used by Puebloans for religious rituals and political meetings.

Late Antique Little Ice Age (LALIA): A strikingly cold period from 536 until ca. 660 CE that enveloped the entire Eurasian continent.

Latewood: Wood formed in late summer, toward the end of the growing season.

Light rings: Tree rings that have latewood cells with smaller than normal diameters and non-thickened cell walls.

Limiting factor: An environmental factor that determines year-to-year variability in tree growth.

Little Ice Age: A relatively cool period (ca. 1500–1850 CE) that followed the Medieval Climate Anomaly and preceded the period of anthropogenic global warming.

Long Count calendar: A nonrepeating, base-20 calendar used by various pre-Columbian Mesoamerican cultures, including the Maya.

Maximum latewood density: The maximum density of the latewood portion of a tree ring, which reflects how much the cell walls in that ring have thickened by the end of the growing season.

Medieval Climate Anomaly: A relatively warm period (ca. 900–1250 CE) in European climate history, concentrated in the North Atlantic region.

Medieval Warm Period: Original, Eurocentric term for the Medieval Climate Anomaly.

Mesolithic: The middle part of the Stone Age, between the Paleolithic and the Neolithic, spanning the period from ca. 13,000 BCE to 3000 BCE in Europe. See also **Neolithic** and **Paleolithic.**

Missing ring: During extremely dry years, some trees skip forming a ring and thus miss a ring for that year. Missing rings can be detected through crossdating.

Moai: Large, monolithic human statues carved out of volcanic tuff by the Rapa Nui people on Easter Island between ca. 1400 and 1680 CE.

Modern Period: A period (ca. 1500–1800) in European history following the Middle Ages and preceding the Industrial Revolution.

Modified Mercalli scale: Seismic-intensity scale that measures the intensity of shaking caused by an earthquake without instrumental measurements. The Modified scale is an improved version of the original Mercalli intensity scale, developed in 1902.

Moraine: The dirt and rocks pushed along by a glacier as it moves forward.

Neolithic: The final, most recent part of the Stone Age, starting ca. 6000 BCE.

Nilometer: A structure (a column, a stairway, or a well with culvert) built to measure the Nile River's water level during its annual flood season.

North Atlantic Oscillation (NAO): A seesaw (or oscillation) in the atmospheric pressure (or air pressure) between two major pressure centers over the North Atlantic Ocean: the Azores High and the Icelandic Low.

Orbital variations: Variations in eccentricity, axial tilt, and precession of Earth's orbit that influence Earth's climate over periods of 100,000, 40,000, and 20,000 years.

Otolith: Bone in the inner ear of vertebrates. Sclerochronologists can count and crossdate the annual growth rings in fish otoliths to extract paleoclimate information. See also **sclerochronology.**

Paleodendrochronology: The study of tree rings in petrified wood.

Paleolithic: Early Stone Age, from the earliest use of stone tools by hominins ca. 3.3 million years BCE to the start of the Mesolithic ca. 13,000 BCE.

Paleotempestology: The study of past storms and tropical cyclones.

Petrified wood: Fossil wood in which all organic material has been replaced by mineral deposits and the original structure of the wood has been maintained.

Pile dwellings: Small Neolithic (ca. 5000–500 BCE) residences constructed on top of posts or piles in lake or bog wetlands.

Pioneer trees: Fast-growing tree species that are often the first to colonize open spaces.

Pluvial: A multi-year period marked by high rainfall.

Pointer year: A year when tree rings in most trees within a region are abnormally narrow or wide.

Polar vortex: An area of low pressure and cold air that surrounds polar regions. When this region expands in winter, midlatitude regions can experience extremely cold temperatures.

Proxy climate records: Natural or man-made archives that record past climate conditions and can therefore be used as sources of climate information.

Radiocarbon dating: A method to determine the age of an organic object by measuring the amount of radiocarbon it detains. Also called *carbon dating* or *carbon-14 dating.*

Reference tree-ring chronology: An absolutely and exactly dated tree-ring chronology, often with high contributing sample replication, that can be used as a reference to crossdate undated tree-ring series from the same region.

Reforestation: The (natural or man-made) development of new forest on land that has been forested before.

Richter scale: A numerical classification of the magnitude of earthquakes based on the strength of their seismic waves.

Ridiculously Resilient Ridge: The nickname for the persistent anticyclone in the eastern North Pacific Ocean that was related to the 2012–16 California drought. See also **anticyclone.**

Ring-porous wood: A type of (hard)wood that is characterized by larger vessels in the earlywood than in the latewood, such as oak wood.

Roman Climate Optimum: A relatively warm period (ca. 300 BCE–200 CE) in Europe and the North Atlantic region. Also called *Roman Warm Period.*

Roman Transition Period: A 300-year period (ca. 250–550 CE) during which the Western Roman Empire transitioned from a sociopolitically complex state to an amalgam of rump states.

Sclerochronology: The study of annual and seasonal growth patterns in the hard tissues of marine organisms, such as mollusk shells, corals, and fish otoliths.

Sinker wood: Sunken logs on the bottom of rivers and lakes.

Smokey Bear effect: Nickname for the effect of large-scale twentieth-century fire suppression on fire regimes in the western US. By persistently preventing frequent surface fires, which are a natural part of the forest ecosystem, the Smokey Bear effect created a century-long fire deficit that has resulted in massive, destructive stand-replacing fires. See also **Fire deficit.**

Snag: Standing dead tree.

Snow Water Equivalent (SWE): A commonly used snowpack measurement that reflects the amount of water contained within the snowpack.

Solar flare: An eruption of intense radiation from the sun's surface that causes electromagnetic disturbances on Earth.

Solar radiation management (SRM): A type of climate engineering in which solar radiation is reflected before it reaches the earth's surface, for instance, by artificially injecting sulfate aerosols into the stratosphere and thus mimicking the effect of volcanic eruptions. See also **Climate engineering.**

Soot: An airborne contaminant consisting of impure carbon particles resulting from the incomplete combustion of hydrocarbons, such as from coal burning, internal-combustion engines, forest fires, and waste incineration. Also called *black carbon.*

Spiral growth: The natural, helical growth form along the trunk of some (often old) trees. Also called *spiral grain.*

Stratosphere: The layer of Earth's atmosphere just above the troposphere but below the mesosphere (ca. 6–30 miles above Earth's surface).

Strip barking: Tree morphology, often found in old trees, in which only a portion, or *strip*, of the stem contains living tissue in the form of cambium.

Subfossil: Partially (rather than fully) fossilized; either not enough time has elapsed yet or preservation conditions are suboptimal for full fossilization.

Sunspots: Regions of reduced temperature on the sun's surface that indicate magnetic activity and appear as patches darker than their surroundings.

Superflare: A very strong solar flare with energies up to ten thousand times that of typical solar flares. See also **Solar flare.**

Surface fire: Low-intensity, nondestructive wildfire that burns on the surface of the ground. Also called *groundfire* or *low-intensity fire*.

Teleconnection: A causal relation between climatic phenomena that occur at large distances, often thousands of miles, apart.

Terminal Classic Period: The final phase (ca. 800–950 CE) of the Classic Maya Period.

Time series: A sequence of data recorded at successive points in time and listed chronologically.

Tree-harvest date: Date of the outer tree ring in an archeological wood sample; gives an indication of the year when the tree was felled, or harvested. Also called *felling date*.

Tree-ring chronology: A time series based on crossdated tree-ring data derived from multiple trees and/or sites.

Tree-ring series: A time series based on tree-ring data derived from a single tree-ring sample.

Tree-ring signature: A distinctive, recognizable sequence of consecutive narrow and wide rings in a regional tree-ring pattern.

Troposphere: The layer of Earth's atmosphere just above Earth's surface, up to ca. 6 miles.

Vessel: Large, tube-like, water-conducting cells in the wood of broadleaf trees.

Völkerwanderung: The period of migration (ca. 250–410 CE) during which widespread migration of Germanic tribes and Huns into Roman territory contributed to the decline of the Western Roman Empire. Also called *Migration Period*.

Bibliography

Prologue

Čufar, K., Beuting, M., Demšar, B., and Merela, M. 2017. Dating of violins—The interpretation of dendrochronological reports. *Journal of Cultural Heritage* 27, S44–S54.

One **Trees in the Desert**

Douglass, A. E. 1914. A method of estimating rainfall by the growth of trees. *Bulletin of the American Geographical Society* 46 (5), 321–35.

Douglass, A. E. 1917. Climatic records in the trunks of trees. *American Forestry* 23 (288), 732–35.

Douglass, A. E. 1929. The secret of the Southwest solved with talkative tree rings. *National Geographic*, December, 736–70.

Hawley, F., Wedel, W. M., and Workman, E. J. 1941. *Tree-ring analysis and dating in the Mississippi drainage*. Chicago: University of Chicago Press.

Lockyer, J. N., and Lockyer, W. J. L. 1901. On solar changes of temperature and variations in rainfall in the region surrounding the Indian Ocean. *Proceedings of the Royal Society of London* 67, 409–31.

Lowell, P. 1895. Mars: The canals I. *Popular Astronomy* 2, 255–61.

Swetnam, T. W., and Brown, P. M. 1992. Oldest known conifers in the southwestern United States: Temporal and spatial patterns of maximum age. *Old growth forests in the Southwest and Rocky Mountain regions*. USDA Forest Service General Technical Report RM-213, 24–38. Fort Collins, CO: USDA Forest Service.

Webb, G. E. 1983. *Tree rings and telescopes: The scientific career of A. E. Douglass*. Tucson: University of Arizona Press.

Two **I Count the Rings Down in Africa**

Dawson, A., Austin, D., Walker, D., Appleton, S., Gillanders, B. M., Griffin, S. M., Sakata, C., and Trouet, V. 2015. A tree-ring based reconstruction of early summer precipitation in southwestern Virginia (1750–1981). *Climate Research* 64 (3), 243–56.

Fritts, H. C. 1976. *Tree rings and climate*. London: Academic.

Trouet, V., Haneca, K., Coppin, P., and Beeckman, H. 2001. Tree ring analysis of Brachystegia spiciformis and Isoberlinia tomentosa: Evaluation of the ENSO-signal in the miombo woodland of eastern Africa. *IAWA Journal* 22 (4), 385–99.

Three Adonis, Methuselah, and Prometheus

Bevan-Jones, R. 2002. *The ancient yew: A history of* Taxus baccata. Macclesfield, Cheshire: Windgather.

Brandes, R. 2007. *Waldgrenzen griechischer Hochgebirge: Unter besonderer Berücksichtigung des Taygetos, Südpeloponnes (Walddynamik, Tannensterben, Dendrochronologie)*. Erlangen-Nürnberg: Friedrich Alexander Universität.

Ferguson, C. W. 1968. Bristlecone pine: Science and esthetics: A 7100-year tree-ring chronology aids scientists; old trees draw visitors to California mountains. *Science* 159 (3817), 839–46.

Klippel, L., Krusic, P. J., Konter, O., St. George, S., Trouet, V., and Esper, J. 2019. A 1200+ year reconstruction of temperature extremes for the northeastern Mediterranean region. *International Journal of Climatology* 39 (4), 2336–50.

Konter, O., Krusic, P. J., Trouet, V., and Esper, J. 2017. Meet Adonis, Europe's oldest dendrochronologically dated tree. *Dendrochronologia* 42, 12.

Stahle, D. W., Edmondson, J. R., Howard, I. M., Robbins, C. R., Griffin, R. D., Carl, A., Hall, C. B., Stahle, D. K., and Torbenson, M. C. A. 2019. Longevity, climate sensitivity, and conservation status of wetland trees at Black River, North Carolina. *Environmental Research Communications* 1 (4), 041002.

Four And the Tree Was Happy

Berlage, H. P. 1931. On the relationship between thickness of tree rings of Djati (teak) trees and rainfall on Java. *Tectona* 24, 939–53.

Bryson, R. A., and Murray, T. 1977. *Climates of hunger: Mankind and the world's changing weather*. Madison: University of Wisconsin Press.

De Micco, V., Campelo, F., De Luis, M., Bräuning, A., Grabner, M., Battipaglia, G., and Cherubini, P. 2016. Intra-annual density fluctuations in tree rings: How, when, where, and why. *IAWA Journal* 37 (2), 232–59.

Francis, J. E. 1986. Growth rings in Cretaceous and Tertiary wood from Antarctica and their palaeoclimatic implications. *Palaeontology* 29 (4), 665–84.

Friedrich, M., Remmele, S., Kromer, B., Hofmann, J., Spurk, M., Kaiser, K. F., Orcel, C., and Küppers, M. 2004. The 12,460-year Hohenheim oak and pine tree-ring chronology from central Europe—A unique annual record for radiocarbon calibration and paleoenvironment reconstructions. *Radiocarbon* 46 (3), 1111–22.

Pilcher, J. R., Baillie, M. G., Schmidt, B., and Becker, B. 1984. A 7,272-year tree-ring chronology for western Europe. *Nature* 312 (5990), 150.

Schöngart, J., Piedade, M. T. F., Wittmann, F., Junk, W. J., and Worbes, M. 2005. Wood growth patterns of *Macrolobium acaciifolium* (Benth.) Benth. (Fabaceae) in Amazonian black-water and white-water floodplain forests. *Oecologia* 145 (3), 454–61.

Silverstein, S., Freeman, N., and Kennedy, A. P. 1964. *The giving tree*. New York: Harper & Row.

Five The Stone Age, the Plague, and Shipwrecks under the City

Billamboz, A. 2004. Dendrochronology in lake-dwelling research. In *Living on the lake in prehistoric Europe: 150 years of lake-dwelling research*, edited by F. Menotti, 117–31. New York: Routledge.

Büntgen, U., Tegel, W., Nicolussi, K., McCormick, M., Frank, D., Trouet, V., Kaplan, J. O., Herzig, F., Heussner, K. U., Wanner, H., Luterbacher, J., and Esper, J. 2011. 2500 years of European climate variability and human susceptibility. *Science* 331 (6017), 578–82.

Daly, A. 2007. The Karschau ship, Schleswig Holstein: Dendrochronological results and timber provenance. *International Journal of Nautical Archaeology* 36 (1), 155–66.

Haneca, K., Wazny, T., Van Acker, J., and Beeckman, H. 2005. Provenancing Baltic timber from art historical objects: Success and limitations. *Journal of Archaeological Science* 32 (2), 261–71.

Hillam, J., Groves, C. M., Brown, D. M., Baillie, M. G. L., Coles, J. M., and Coles, B. J. 1990. Dendrochronology of the English Neolithic. *Antiquity* 64 (243), 210–20.

Martin-Benito, D., Pederson, N., McDonald, M., Krusic, P., Fernandez, J. M., Buckley, B., Anchukaitis, K. J., D'Arrigo, R., Andreu-Hayles, L., and Cook, E. 2014. Dendrochronological dating of the World Trade Center ship, Lower Manhattan, New York City. *Tree-Ring Research* 70 (2), 65–77.

Miles, D. W. H., and Bridge, M. C. 2005. *The tree-ring dating of the early medieval doors at Westminster Abbey, London*. English Heritage Centre for Archaeology, Report 38/2005. London: English Heritage.

Pearson, C. L., Brewer, P. W., Brown, D., Heaton, T. J., Hodgins, G. W., Jull, A. T., Lange, T., and Salzer, M. W. 2018. Annual radiocarbon record indicates 16th century BCE date for the Thera eruption. *Science Advances* 4 (8), eaar8241.

Reimer, P. J., Bard, E., Bayliss, A., Beck, J. W., Blackwell, P. G., Ramsey, C. B., Buck, C. E., Cheng, H., Edwards, R. L., Friedrich, M., and Grootes, P. M. 2013. IntCal13 and Marine13 radiocarbon age calibration curves 0–50,000 years cal BP. *Radiocarbon* 55 (4), 1869–87.

Slayton, J. D., Stevens, M. R., Grissino-Mayer, H. D., and Faulkner, C. H. 2009. The historical dendroarchaeology of two log structures at the Marble Springs Historic Site, Knox County, Tennessee, USA. *Tree-Ring Research* 65 (1), 23–36.

Tegel, W., Elburg, R., Hakelberg, D., Stäuble, H., and Büntgen, U. 2012. Early Neolithic water wells reveal the world's oldest wood architecture. *PloS One* 7 (12), e51374.

Six **The Hockey Stick Poster Child**

Bradley, R. S. 2011. *Global warming and political intimidation: How politicians cracked down on scientists as the earth heated up.* Amherst: University of Massachusetts Press.

Büntgen, U., Frank, D., Grudd, H., and Esper, J. 2008. Long-term summer temperature variations in the Pyrenees. *Climate Dynamics* 31 (6), 615–31.

Büntgen, U., Frank, D., Trouet, V., and Esper, J. 2010. Diverse climate sensitivity of Mediterranean tree-ring width and density. *Trees* 24 (2), 261–73.

Mann, M. E., Bradley, R. S., and Hughes, M. K. 1998. Global-scale temperature patterns and climate forcing over the past six centuries. *Nature* 392 (6678), 779.

Mann, M. E., Bradley, R. S., and Hughes, M. K. 1999. Northern Hemisphere temperatures during the past millennium: Inferences, uncertainties, and limitations. *Geophysical Research Letters* 26 (6), 759–62.

Oreskes, N., and Conway, E. M. 2011. *Merchants of doubt: How a handful of scientists obscured the truth on issues from tobacco smoke to global warming.* New York: Bloomsbury.

Seven **Wind of Change**

Esper, J., Frank, D., Büntgen, U., Verstege, A., Luterbacher, J., and Xoplaki, E. 2007. Long-term drought severity variations in Morocco. *Geophysical Research Letters* 34 (17), L07711.

Frank, D. C., Esper, J., Raible, C. C., Büntgen, U., Trouet, V., Stocker, B., and Joos, F. 2010. Ensemble reconstruction constraints on the global carbon cycle sensitivity to climate. *Nature* 463 (7280), 527.

Frank, D. C., Esper, J., Zorita, E., and Wilson, R. 2010. A noodle, hockey stick, and spaghetti plate: A perspective on high-resolution paleoclimatology. *Wiley Interdisciplinary Reviews: Climate Change* 1 (4), 507–16.

Lamb, H. H. 1965. The early medieval warm epoch and its sequel. *Palaeogeography, Palaeoclimatology, Palaeoecology* 1, 13–37.

Proctor, C. J., Baker, A., Barnes, W. L., and Gilmour, M. A. 2000. A thousand year speleothem proxy record of North Atlantic climate from Scotland. *Climate Dynamics* 16 (10–11), 815–20.

Trouet, V., Esper, J., Graham, N. E., Baker, A., Scourse, J. D., and Frank, D. C. 2009. Persistent positive North Atlantic Oscillation mode dominated the medieval climate anomaly. *Science* 324 (5923), 78–80.

Eight Winter Is Coming

Brázdil, R., Kiss, A., Luterbacher, J., Nash, D. J., and Řezníčková, L. 2018. Documentary data and the study of past droughts: A global state of the art. *Climate of the Past* 14 (12), 1915–60.

Degroot, D. 2018. Climate change and society in the 15th to 18th centuries. *Wiley Interdisciplinary Reviews: Climate Change* 9 (3), e518.

Fagan, B. 2000. *The Little Ice Age*. New York: Basic Books.

Le Roy, M., Nicolussi, K., Deline, P., Astrade, L., Edouard, J. L., Miramont, C., and Arnaud, F. 2015. Calendar-dated glacier variations in the western European Alps during the Neoglacial: The Mer de Glace record, Mont Blanc massif. *Quaternary Science Reviews* 108, 1–22.

Le Roy Ladurie, E. 1971. *Times of feast, times of famine: A history of climate since the year 1000*. New York: Doubleday.

Ludlow, F., Stine, A. R., Leahy, P., Murphy, E., Mayewski, P. A., Taylor, D., Killen, J., Baillie, M. G., Hennessy, M., and Kiely, G. 2013. Medieval Irish chronicles reveal persistent volcanic forcing of severe winter cold events, 431–1649 CE. *Environmental Research Letters* 8 (2), 024035.

Magnusson, M., and Pálsson, H. 1965. The Vinland Sagas: Grænlendiga Saga and Eirik's Saga. Harmondsworth: Penguin.

Nelson, M. C., Ingram, S. E., Dugmore, A. J., Streeter, R., Peeples, M. A., McGovern, T. H., Hegmon, M., Arneborg, J., Kintigh, K. W., Brewington, S., and Spielmann, K. A. 2016. Climate challenges, vulnerabilities, and food security. *Proceedings of the National Academy of Sciences* 113 (2), 298–303.

Nine Three Tree-Ring Scientists Walk into a Bar

Belmecheri, S., Babst, F., Wahl, E. R., Stahle, D. W., and Trouet, V. 2016. Multi-century evaluation of Sierra Nevada snowpack. *Nature Climate Change* 6 (1), 2.

Black, B. A., Sydeman, W. J., Frank, D. C., Griffin, D., Stahle, D. W., García-Reyes, M., Rykaczewski, R. R., Bograd, S. J., and Peterson, W. T. 2014. Six centuries of variability and extremes in a coupled marine-terrestrial ecosystem. *Science* 345 (6203), 1498–1502.

Butler, P. G., Wanamaker, A. D., Scourse, J. D., Richardson, C. A., and Reynolds, D. J. 2013. Variability of marine climate on the North Icelandic Shelf in a 1357-year proxy archive based on growth increments in the bivalve *Arctica islandica*. *Palaeogeography, Palaeoclimatology, Palaeoecology* 373, 141–51.

Griffin, D., and Anchukaitis, K. J. 2014. How unusual is the 2012–2014 California drought? *Geophysical Research Letters* 41 (24), 9017–23.

Marx, R. F. 1987. *Shipwrecks in the Americas*. New York: Crown.

Stahle, D. W., Griffin, R. D., Meko, D. M., Therrell, M. D., Edmondson, J. R., Cleaveland, M. K., Stahle, L. N., Burnette, D. J., Abatzoglou, J. T., Redmond, K. T., and Dettinger, M. D. 2013. The ancient blue oak woodlands of California: Longevity and hydroclimatic history. *Earth Interactions* 17 (12), 1–23.

Trouet, V., Harley, G. L., and Domínguez-Delmás, M. 2016. Shipwreck rates reveal Caribbean tropical cyclone response to past radiative forcing. *Proceedings of the National Academy of Sciences* 13 (12), 3169–74.

Ten Ghosts, Orphans, and Extraterrestrials

Atwater, B. F., Musumi-Rokkaku, S., Satake, K., Tsuji, Y., Ueda, K., and Yamaguchi, D. K. 2016. *The orphan tsunami of 1700: Japanese clues to a parent earthquake in North America.* Seattle: University of Washington Press.

Briffa, K. R., Jones, P. D., Schweingruber, F. H., and Osborn, T. J. 1998. Influence of volcanic eruptions on Northern Hemisphere summer temperature over the past 600 years. *Nature* 393 (6684), 450.

LaMarche, V. C., Jr, and Hirschboeck, K. K. 1984. Frost rings in trees as records of major volcanic eruptions. *Nature* 307 (5947), 121.

Manning, J. G., Ludlow, F., Stine, A. R., Boos, W. R., Sigl, M., and Marlon, J. R. 2017. Volcanic suppression of Nile summer flooding triggers revolt and constrains interstate conflict in ancient Egypt. *Nature Communications* 8 (1), 900.

Miyake, F., Nagaya, K., Masuda, K., and Nakamura, T. 2012. A signature of cosmic-ray increase in AD 774–775 from tree rings in Japan. *Nature* 486 (7402), 240.

Mousseau, T. A., Welch, S. M., Chizhevsky, I., Bondarenko, O., Milinevsky, G., Tedeschi, D. J., Bonisoli-Alquati, A., and Møller, A. P. 2013. Tree rings reveal extent of exposure to ionizing radiation in Scots pine *Pinus sylvestris. Trees* 27 (5), 1443–53.

Munoz, S. E., Giosan, L., Therrell, M. D., Remo, J. W., Shen, Z., Sullivan, R. M., Wiman, C., O'Donnell, M., and Donnelly, J. P. 2018. Climatic control of Mississippi River flood hazard amplified by river engineering. *Nature* 556 (7699), 95.

Pang, K. D. 1991. The legacies of eruption: Matching traces of ancient volcanism with chronicles of cold and famine. *The Sciences* 31 (1), 30–35.

Sigl, M., Winstrup, M., McConnell, J. R., Welten, K. C., Plunkett, G., Ludlow, F., Büntgen, U., Caffee, M., Chellman, N., Dahl-Jensen, D., and Fischer, H. 2015. Timing and climate forcing of volcanic eruptions for the past 2,500 years. *Nature* 523 (7562), 543.

Therrell, M. D., and Bialecki, M. B. 2015. A multi-century tree-ring record of spring flooding on the Mississippi River. *Journal of Hydrology* 529, 490–98.

Vaganov, E. A., Hughes, M. K., Silkin, P. P., and Nesvetailo, V. D. 2004. The Tunguska event in 1908: Evidence from tree-ring anatomy. *Astrobiology* 4 (3), 391–99.

Eleven **Disintegration, or The Fall of Rome**

Baker, A., Hellstrom, J. C., Kelly, B. F., Mariethoz, G., and Trouet, V. 2015. A composite annual-resolution stalagmite record of North Atlantic climate over the last three millennia. *Scientific Reports* 5, 10307.

Büntgen, U., Myglan, V. S., Ljungqvist, F. C., McCormick, M., Di Cosmo, N., Sigl, M., Jungclaus, J., Wagner, S., Krusic, P. J., Esper, J., and Kaplan, J. O. 2016. Cooling and societal change during the Late Antique Little Ice Age from 536 to around 660 AD. *Nature Geoscience* 9 (3), 231–36.

Büntgen, U., Tegel, W., Nicolussi, K., McCormick, M., Frank, D., Trouet, V., Kaplan, J. O., Herzig, F., Heussner, K. U., Wanner, H., Luterbacher, J., and Esper, J. 2011. 2500 years of European climate variability and human susceptibility. *Science* 331 (6017), 578–82.

Diaz, H., and Trouet, V. 2014. Some perspectives on societal impacts of past climatic changes. *History Compass* 12 (2), 160–77.

Dull, R. A., Southon, J. R., Kutterolf, S., Anchukaitis, K. J., Freundt, A., Wahl, D. B., Sheets, P., Amaroli, P., Hernandez, W., Wiemann, M. C., and Oppenheimer, C. 2019. Radiocarbon and geologic evidence reveal Ilopango volcano as source of the colossal 'mystery' eruption of 539/40 CE. *Quaternary Science Reviews* 222, 105855.

Harper, K. 2017. *The fate of Rome: Climate, disease, and the end of an empire*. Princeton, NJ: Princeton University Press.

Helama, S., Arppe, L., Uusitalo, J., Holopainen, J., Mäkelä, H. M., Mäkinen, H., Mielikäinen, K., Nöjd, P., Sutinen, R., Taavitsainen, J. P., and Timonen, M. 2018. Volcanic dust veils from sixth century tree-ring isotopes linked to reduced irradiance, primary production and human health. *Scientific Reports* 8 (1), 1339.

Sheppard, P. R., Tarasov, P. E., Graumlich, L. J., Heussner, K. U., Wagner, M., Österle, H., and Thompson, L. G. 2004. Annual precipitation since 515 BC reconstructed from living and fossil juniper growth of northeastern Qinghai Province, China. *Climate Dynamics* 23 (7–8), 869–81.

Soren, D. 2002. *Malaria, witchcraft, infant cemeteries, and the fall of Rome*. San Diego: Department of Classics and Humanities, San Diego State University.

Soren, D. 2003. Can archaeologists excavate evidence of malaria? *World Archaeology* 35 (2), 193–209.

Stothers, R. B., and Rampino, M. R. 1983. Volcanic eruptions in the Mediterranean before AD 630 from written and archaeological sources. *Journal of Geophysical Research: Solid Earth* 88 (B8), 6357–71.

Twelve **It's the End of the World as We Know It**

Acuna-Soto, R., Stahle, D. W., Therrell, M. D., Chavez, S. G., and Cleaveland, M. K. 2005. Drought, epidemic disease, and the fall of classic period cultures in Meso-america (AD 750–950): Hemorrhagic fevers as a cause of massive population loss. *Medical Hypotheses* 65 (2), 405–9.

Buckley, B. M., Anchukaitis, K. J., Penny, D., Fletcher, R., Cook, E. R., Sano, M., Wichienkeeo, A., Minh, T. T., and Hong, T. M. 2010. Climate as a contributing factor in the demise of Angkor, Cambodia. *Proceedings of the National Academy of Sciences* 107 (15), 6748–52.

Di Cosmo, N., Hessl, A., Leland, C., Byambasuren, O., Tian, H., Nachin, B., Pederson, N., Andreu-Hayles, L., and Cook, E. R. 2018. Environmental stress and steppe nomads: Rethinking the history of the Uyghur Empire (744–840) with paleoclimate data. *Journal of Interdisciplinary History* 48 (4), 439–63.

Hessl, A. E., Anchukaitis, K. J., Jelsema, C., Cook, B., Byambasuren, O., Leland, C., Nachin, B., Pederson, N., Tian, H., and Hayles, L. A. 2018. Past and future drought in Mongolia. *Science Advances* 4 (3), e1701832.

Huntington, E. 1917. Maya civilization and climate changes. Paper presented at the XIX International Congress of Americanists, Washington, DC.

Pederson, N., Hessl, A. E., Baatarbileg, N., Anchukaitis, K. J., and Di Cosmo, N. 2014. Pluvials, droughts, the Mongol Empire, and modern Mongolia. *Proceedings of the National Academy of Sciences* 111 (12), 4375–79.

Sano, M., Buckley, B. M., and Sweda, T. 2009. Tree-ring based hydroclimate reconstruction over northern Vietnam from *Fokienia hodginsii*: Eighteenth century mega-drought and tropical Pacific influence. *Climate Dynamics* 33 (2–3), 331.

Stahle, D. W., Diaz, J. V., Burnette, D. J., Paredes, J. C., Heim, R. R., Fye, F. K., Soto, R. A., Therrell, M. D., Cleaveland, M. K., and Stahle, D. K. 2011. Major Mesoamerican droughts of the past millennium. *Geophysical Research Letters* 38, L05703.

Therrell, M. D., Stahle, D. W., and Acuna-Soto, R. 2004. Aztec drought and the "curse of one rabbit." *Bulletin of the American Meteorological Society* 85 (9), 1263–72.

Thirteen **Once upon a Time in the West**

American Association for the Advancement of Science. 1921. The Pueblo Bonito expedition of the National Geographic Society. *Science* 54 (1402), 458.

Bocinsky, R. K., Rush, J., Kintigh, K. W., and Kohler, T. A. 2016. Exploration and exploitation in the macrohistory of the pre-Hispanic Pueblo Southwest. *Science Advances* 2 (4), e1501532.

Cook, E. R., Woodhouse, C. A., Eakin, C. M., Meko, D. M., and Stahle, D. W. 2004. Long-term aridity changes in the western United States. *Science* 306 (5698), 1015–18.

Dean, J. S. 1967. *Chronological analysis of Tsegi phase sites in northeastern Arizona.* Papers of the Laboratory of Tree-Ring Research, No. 3. Tucson: University of Arizona Press.

Dean, J. S., and Warren, R. L. 1983. Dendrochronology. In *The architecture and dendrochronology of Chetro Ketl*, edited by S. H. Lekson, 105–240. Reports of the Chaco Center, No. 6. Albuquerque: National Park Service.

Douglass, A. E. 1935. *Dating Pueblo Bonito and other ruins of the Southwest*. Pueblo Bonito Series, No. 1. Washington, DC: National Geographic Society.

Frazier, K. 1999. *People of Chaco: A canyon and its culture*. New York: Norton.

Guiterman, C. H., Swetnam, T. W., and Dean, J. S. 2016. Eleventh-century shift in timber procurement areas for the great houses of Chaco Canyon. *Proceedings of the National Academy of Sciences* 113 (5), 1186–90.

Meko, D. M., Woodhouse, C. A., Baisan, C. A., Knight, T., Lukas, J. J., Hughes, M. K., and Salzer, M. W. 2007. Medieval drought in the upper Colorado River basin. *Geophysical Research Letters* 34 (10), L10705.

Stahle, D. W., Cleaveland, M. K., Grissino-Mayer, H. D., Griffin, R. D., Fye, F. K., Therrell, M. D., Burnette, D. J., Meko, D. M., and Villanueva Diaz, J. 2009. Cool- and warm-season precipitation reconstructions over western New Mexico. *Journal of Climate* 22 (13), 3729–50.

Stockton, C. W., and Jacoby, G. C. 1976. *Long-term surface water supply and streamflow trends in the Upper Colorado River basin*. Lake Powell Research Project Bulletin No. 18. Arlington, VA: National Science Foundation.

Windes, T. C., and McKenna, P. J. 2001. Going against the grain: Wood production in Chacoan society. *American Antiquity* 66 (1), 119–40.

Woodhouse, C. A., Meko, D. M., MacDonald, G. M., Stahle, D. W., and Cook, E. R. 2010. A 1,200-year perspective of 21st century drought in southwestern North America. *Proceedings of the National Academy of Sciences* 107 (50), 21283–88.

Fourteen **Will the Wind Ever Remember?**

Alfaro-Sánchez, R., Nguyen, H., Klesse, S., Hudson, A., Belmecheri, S., Köse, N., Diaz, H. F., Monson, R. K., Villalba, R., and Trouet, V. 2018. Climatic and volcanic forcing of tropical belt northern boundary over the past 800 years. *Nature Geoscience* 1 (12), 933–38.

Cook, B. I., Williams, A. P., Mankin, J. S., Seager, R., Smerdon, J. E., and Singh, D. 2018. Revisiting the leading drivers of Pacific coastal drought variability in the contiguous United States. *Journal of Climate* 31 (1), 25–43.

Fang, J. Q. 1992. Establishment of a data bank from records of climatic disasters and anomalies in ancient Chinese documents. *International Journal of Climatology* 12 (5), 499–519.

Li, J., Xie, S. P., Cook, E. R., Morales, M. S., Christie, D. A., Johnson, N. C., Chen, F., D'Arrigo, R., Fowler, A. M., Gou, X, and Fang, K. 2013. El Niño modulations over the past seven centuries. *Nature Climate Change* 3 (9), 822.

Shen, C., Wang, W. C., Hao, Z., and Gong, W. 2007. Exceptional drought events over eastern China during the last five centuries. *Climatic Change* 85 (3–4), 453–71.

Stahle, D. W., Cleaveland, M. K., Blanton, D. B., Therrell, M. D., and Gay, D. A. 1998. The lost colony and Jamestown droughts. *Science* 280 (5363), 564–67.

Trouet, V., Babst, F., and Meko, M. 2018. Recent enhanced high-summer North Atlantic Jet variability emerges from three-century context. *Nature Communications* 9 (1), 180.

Trouet, V., Panayotov, M. P., Ivanova, A., and Frank, D. 2012. A pan-European summer teleconnection mode recorded by a new temperature reconstruction from the northeastern Mediterranean (AD 1768–2008). *Holocene* 22 (8), 887–98.

Urban, F. E., Cole, J. E., and Overpeck, J. T. 2000. Influence of mean climate change on climate variability from a 155-year tropical Pacific coral record. *Nature* 407 (6807), 989.

White, S. 2011. *The climate of rebellion in the early modern Ottoman Empire*. Cambridge: Cambridge University Press.

Fifteen **After the Gold Rush**

Abatzoglou, J. T., and Williams, A. P. 2016. Impact of anthropogenic climate change on wildfire across western US forests. *Proceedings of the National Academy of Sciences* 113 (42), 11770–75.

Anderson, K. 2005. *Tending the wild: Native American knowledge and the management of California's natural resources*. Berkeley: University of California Press.

Dennison, P. E., Brewer, S. C., Arnold, J. D., and Moritz, M. A. 2014. Large wildfire trends in the western United States, 1984–2011. *Geophysical Research Letters* 41 (8), 2928–33.

Fenn, E. A. 2001. *Pox Americana: The great smallpox epidemic of 1775–82*. New York: Hill & Wang.

Liebmann, M. J., Farella, J., Roos, C. I., Stack, A., Martini, S., and Swetnam, T. W. 2016. Native American depopulation, reforestation, and fire regimes in the Southwest United States, 1492–1900 CE. *Proceedings of the National Academy of Sciences* 113 (6), E696–E704.

Muir, J. 1961. *The mountains of California*. 1894. Reprint. New York: American Museum of Natural History and Doubleday.

Swetnam, T. W. 1993. Fire history and climate change in giant sequoia groves. *Science* 262 (5135), 885–89.

Swetnam, T. W., and Betancourt, J. L. 1990. Fire–southern oscillation relations in the southwestern United States. *Science* 249 (4972), 1017–20.

Taylor, A. H., Trouet, V., Skinner, C. N., and Stephens, S. 2016. Socioecological transitions trigger fire regime shifts and modulate fire-climate interactions in the

Sierra Nevada, USA, 1600–2015 CE. *Proceedings of the National Academy of Sciences* 113 (48), 13684–89.

Trouet, V., Taylor, A. H., Wahl, E. R., Skinner, C. N., and Stephens, S. L. 2010. Fire-climate interactions in the American West since 1400 CE. *Geophysical Research Letters* 37 (4), L18704.

Westerling, A. L., Hidalgo, H. G., Cayan, D. R., and Swetnam, T. W. 2006. Warming and earlier spring increase western US forest wildfire activity. *Science* 313 (5789), 940–43.

Sixteen **The Forest for the Trees**

Appuhn, K. 2009. *A forest on the sea: Environmental expertise in Renaissance Venice.* Baltimore: Johns Hopkins University Press.

Babst, F., Alexander, M. R., Szejner, P., Bouriaud, O., Klesse, S., Roden, J., Ciais, P., Poulter, B., Frank, D., Moore, D. J., and Trouet, V. 2014. A tree-ring perspective on the terrestrial carbon cycle. *Oecologia* 176 (2), 307–22.

Conard, N. J., Serangeli, J., Böhner, U., Starkovich, B. M., Miller, C. E., Urban, B., and Van Kolfschoten, T. 2015. Excavations at Schöningen and paradigm shifts in human evolution. *Journal of Human Evolution* 89, 1–17.

Corcoran, P. L., Moore, C. J., and Jazvac, K. 2014. An anthropogenic marker horizon in the future rock record. *GSA Today* 24 (6), 4–8.

Flenley, J. R., and King, S. M. 1984. Late quaternary pollen records from Easter Island. *Nature* 307 (5946), 47–50.

Hunt, T. L., and Lipo, C. P. 2006. Late colonization of Easter Island. *Science* 311 (5767), 1603–6.

Kaplan, J. O., Krumhardt, K. M., and Zimmermann, N. E. 2012. The effects of land use and climate change on the carbon cycle of Europe over the past 500 years. *Global Change Biology* 18 (3), 902–14.

Ruddiman, W. F. 2010. *Plows, plagues, and petroleum: How humans took control of climate.* Princeton, NJ: Princeton University Press.

Sandars, N. 1972. *The epic of Gilgamesh.* London: Penguin.

Thieme, H. 1997. Lower Palaeolithic hunting spears from Germany. *Nature* 385 (6619), 807.

Turney, C. S., Palmer, J., Maslin, M. A., Hogg, A., Fogwill, C. J., Southon, J., Fenwick, P., Helle, G., Wilmshurst, J. M., McGlone, M., and Ramsey, C. B. 2018. Global peak in atmospheric radiocarbon provides a potential definition for the onset of the Anthropocene epoch in 1965. *Scientific Reports* 8 (1), 3293.

Acknowledgments

I am immensely grateful to a great many people who have helped me to turn this book from idea into reality. First and foremost, I want to thank Tiffany Gasbarrini, my editor at Johns Hopkins University Press, for planting the idea in my head and for helping me to bring it to fruition. Thank you, Tiffany, for your *incredible* enthusiasm, for keeping me on track, and for dragging me across the line. Thank you also to Joanne Allen, Esther Rodriguez, and the whole JHUP team for taking care of every little detail that was involved in publishing this book. I also want to thank Oliver Uberti for enthusiastically hopping on board and for bringing your finest data-visualization ideas to the table.

Thank you to everyone who shared their tree stories with me. I thank especially Jeff Dean, Amy Hessl, Fritz Schweingruber, Dave Stahle, and Ron Towner for letting me ask very many questions. Thanks also go to Chris Baisan, Soumaya Belmecheri, Ulf Büntgen, Jan Esper, David Frank, Kristof Haneca, Claudia Hartl, Paul Krusic, and Tom Swetnam for letting me tell unabridged versions of our stories.

Many thanks to all formal and informal reviewers, who greatly helped improve the manuscript along the way. I thank Kym Coco, Dagomar Degroot, Luc Delesie, Henry Diaz, Paul Krusic, Peter Kuniholm, Jennifer Mix, Neil Pederson, Randall Smith, Pieter Zuidema, and three anonymous reviewers. Many thanks also to Brian Atwater, Mike Baillie, Kyle Bocinsky, Brendan Buckley, Paolo Cherubini, Ed Cook, Holger Gärtner, Chris Guiterman, Zakia Hassan Khamisi, Malcolm Hughes, Melaine Le Roy, Le Canh Nam, Scott Nichols, Charlotte Pearson, Tom Swetnam, Alan Taylor, Willy Tegel, Matt Therrell, and Ed Wright for sharing data, photos, and/or references with me.

I am grateful for the support that I received for my research from the National Science Foundation (grant AGS-1349942) and for this book from the

University of Arizona Udall Center Fellows Program and from the UA Provost's Author Support Fund. I am also grateful to my research group—Raquel Alfaro-Sánchez, Tom De Mil, Amy Hudson, Matt Meko, Guobao Xu, Diana Zamora-Reyes—and to my collaborators for their patience and independence while I was working on the book.

Thank you to all my friends who have let me discuss *Tree Story* whenever I wanted to, for however long I wanted to. Thank you to Erica Bigio, Nathalie Carpentier, Els De Gersem, Bart Eeckhout, Andrea Finger, Rachel Gallery, Moira Heyn, Kris Kuppens, David Moore, Tom Spittaels, Simone Stopford, and Corien Van Zweden. Thanks to my family, my mom and my sisters, for their enduring support. Most of all, I thank Wil Peterson, for giving me a home, a writing nook, and all.

Index

diversity of applications, 17; origins and evolution, 7; "Rosetta Stone" of, 15, 16, 158–59
dendroclimatology, 26
dendrogeomorphology, 94–95
dendroprovenancing, 2–3, 62, 154
deserts, 7, 8, 9, 172
Diaz, Jose Villaneuva, 148, 149, 150
diseases, 149, 150
Domínguez-Delmás, Marta, 107–8
Douglas fir, 153–54
Douglass, Andrew Ellicott, 38, 45, 48, 78, 147, 152, 158–59; as astronomer, 8, 9–12, 20; as dendrochronology founder, 11–17, 18, 20
drought reconstructions, 27–28; American Southwest, 158–64, 177; Asia, 132–33, 145–46; California, 5, 100–105, 147–48, 160, 162, 170; Europe, 86–89; mega-droughts, 158–61; Meso-America, 148–50; Mongolia, 140–42, 143–44; Morocco, 82–84, 86, 88–89; North America, 177; Roman Empire, 131–32; tropics expansion and, 173, 175–77, 178
droughts, 44–45, 92, 100, 206; and El Niño Southern Oscillation; effect on crossdating, 50, 51;, 178–79; snow droughts, 101–5; 20-year, 162, 165; as wildfire risk factor, 186–87, 189–90, 193
Dutch Republic, 96
dzud, 144

earlywood, 34–35, 42, 152, 159–60
earthquakes, 113–16
Easter Island (Rapa Nui), 201–2
Egypt, 126–28, 137
Ellis, Florence Hawley, 20–21
El Niño Southern Oscillation (ENSO), 99, 165, 188, 190–91; El Niño phase, 84–85, 178–79, 188; La Niña phase, 162, 178, 179
enhanced greenhouse effect, 72, 74, 76–77, 78, 111, 165, 180, 204–5; counterbalancing of, 207–8; expansion of the tropics and, 173
epidemics, 149, 150, 176, 177, 192–93, 206–7
Erik the Red, 97
Esper, Jan, 5, 31, 66, 67, 82–83, 84, 88, 90–91
eucalypt, 34
Europe, 207; atmospheric pressure centers, 84–86, 88–91; drought reconstruction,

86–89; jet stream dipole pattern, 167–69; medieval climate change, 93–96; oldest living trees in, 32, 38, 87
extreme weather events, 119–20; jet stream-related, 169–72. See also floods; hurricanes

Fate of Rome: Climate, Disease, and the End of an Empire (Harper), 137
fire-scarred trees, 28, 182–97; cat-face scars, 182–83; in Siberia, 193–97; in Sierra Nevada, 188–93
fir trees, 154
floating-tree chronologies, 14–17, 158–59
flood rings, 119–20
floods, 105–12, 119–20, 206; ENSO-related, 178–79; monsoon, 145–46, 159–60; Nile River, 126–28, 131–32
Florida Keys, 106–9
fossil fuels, 56, 76, 204–5
Four Corners Region. See Ancestral Puebloans
Frank, David, 33, 66, 80, 84, 90–91
Frankenstein (Shelley), 96, 122
frost events, 119–22

Galileo Galilei, 76
gas, 56, 204
Genghis Khan, 140–43
German oak-pine chronology, 6, 51–52
Germanic tribes, 130, 132–33
Germany, 6, 51–52, 55–57
ghost forest trees, 114–16
Giffords, Gabrielle (Gabby) Dee, 133
Gilgamesh, 202–3
girth, of trees, 12, 29, 34
Giving Tree, The (Silverstein), 44
glaciers, 93–95, 205
global climate models, 111–12
global warming, 81–82, 165, 205, 207–8. See also Hockey Stick temperature graph; Spaghetti Plate temperature graph
Gold Rush, 190, 191, 193
Goodall, Jane, 23, 24
Gore, Albert (Al) Arnold, Jr., 71
Great Basin, 35–37
Greece, 30–32
greenhouse effect, 53–54. See also enhanced greenhouse effect